A Photographic Guide to

TREES
OF AUSTRALIA

Denise Greig

Dedicated to my mother, who has, for many years, been an extremely patient, tolerant and helpful assistant and companion during our journeys around this vast country photographing the plants of Australia.

Published in Australia by
New Holland Publishers (Australia) Pty Ltd
Sydney • Auckland • London • Cape Town

14 Aquatic Drive Frenchs Forest NSW 2086 Australia
218 Lake Road Northcote Auckland New Zealand
24 Nutford Place London W1H 5DQ United Kingdom
80 McKenzie Street Cape Town 8001 South Africa

First published in 1998 and reprinted in 2001.

National Library of Australia Cataloguing-in-Publication Data:

Greig, Denise, 1945–.
A photographic guide to trees of Australia.
Includes index.

ISBN 1 86436 326 6.

1. Trees—Australia. 2. Trees—Australia—Identification.
3. Trees—Australia—Pictorial Works. I. Title
(Series: Photoguides).
582.16

Publishing General Manager: Jane Hazell
Publisher: Averill Chase
Project Manager: Fiona Doig
Edited by Lynn Cole
Page layout by DiZign
Reproduction by cmyk pre-press, Cape Town
Printed and bound by Tien Wah Press (Pte) Ltd
Front Cover: River Red Gum; Back Cover: Acacia; Spine: Boab;
Title Page: Strangler Fig

Photographic Acknowledgments
Shaen Adey/New Holland Image Library cover photos and title page. **M. Fagg** p. 26 top, 74 bottom; **M. Fagg/NBG** p. 51 top, 64 bottom, 68, 96 top, 103, 109; **P. Ollerenshaw/NBG** p. 101 top; **E. Phillips/NBG** p. 72 bottom; **J. Wrigley/NBG** p. 126 top. (NBG = National Botanical Gardens.)

Contents

Introduction

Trees mean many things to different people. Gardeners like to grow them for pleasure, shade, beauty or profit; foresters are employed to chop them down to provide us with wood and paper products; artists and poets are inspired by them; bushwalkers admire them; spiritual people hug them and Aborigines use their flowers, foliage and bark for medicine and food. Conservationists protect them, scientists study them, journalists write about them and politicians change their minds about them. Whatever our inclinations, we need trees for the air we breath and the soil we stand on. They are the basis of all life.

Roughly, a tree is defined as a perennial plant with a distinct, self-supporting woody trunk. In most cases, the trunk does not bear branches for some considerable distance above the ground.

In many parts of Australia, trees are the most conspicuous and familiar plants. The aim of this book is to present a diverse range of common and widespread trees that are most likely to be seen in accessible or regularly visited places. Some of the species mentioned are an important component of a small geographical area.

Each tree selected is individually treated with one or more colour photographs, descriptive information, habitat notes and a distribution map. Aboriginal plant usage is also recorded, where applicable.

The enjoyment of trees is partly in seeing them and partly in being able to put a name to those you encounter. The real joy, however, is knowing that we will see them again. The obvious beauty of our trees is matched only by the importance of maintaining and preserving them.

How to use this book

Small illustrations showing characteristics for each family, or different groups within a family, are grouped at the front of the book for easy reference. The same illustrations are shown at the top corner of the pages in the species account and similar species are grouped together.

Headings

Common names are listed where they are known, followed by the most recent scientific name based on current scientific works. The size refers to the height the tree normally attains in maturity. Some individuals may be smaller or larger than the figures shown but most will fall within this range.

The trees described

The trees described and illustrated are the most common or those most likely to be seen in often-frequented places in Australia. With very few exceptions, the species are arranged in an order that is widely accepted by taxonomists and followed by most other books on the plants of Australia. Closely related plant families are grouped together. Genera are arranged alphabetically within each family. Species are also listed alphabetically, except for *Acacia*, which are grouped according to the shape of the flowerheads and leaf types. The eucalypts are roughly divided into two groups: smooth-barked eucalypts and rough-barked eucalypts. The new genus *Corymbia*, which comprises those eucalypts traditionally known as bloodwoods and ghost gums, is recognised.

Unless it is noted that a tree is deciduous, assume that it is evergreen, because most Australian trees are evergreen.

When first attempting to identify a tree in the wild, make sure that the distribution indicated on the map coincides with your own geographical location. Points to look for when identifying a tree are the height and growth habit, bark, leaves, flowers and fruits. Compare these with the illustration and descriptive text provided.

Terminology

Botanical terminology has been kept to a minimum, but in some instances it is unavoidable, so a glossary has been included to assist the reader in understanding some of these terms.

Distribution maps

These maps were compiled from available recorded data. Arrows point to specific locality records; green areas represent broader distribution patterns. This does not imply that a tree is found in every part of that range, only that it is likely to occur in appropriate habitats within the range.

The vegetation of Australia

Travellers in Australia are often surprised at the different varieties of trees seen along the way. Even eucalypts are very variable in appearance, from the contorted Snow Gum, *Eucalyptus pauciflora*, at the inhospitable tree line in alpine regions, to the towering Karri, *E. diversicolor*, of the tall, wet sclerophyll forests of south-western Australia. This great diversity is chiefly a product of the pronounced variation in vegetation types. A tree's preferred vegetation type is mentioned in each description and an understanding of these zones will greatly assist the tree buff or student to discover which trees he can expect to find in a specific zone.

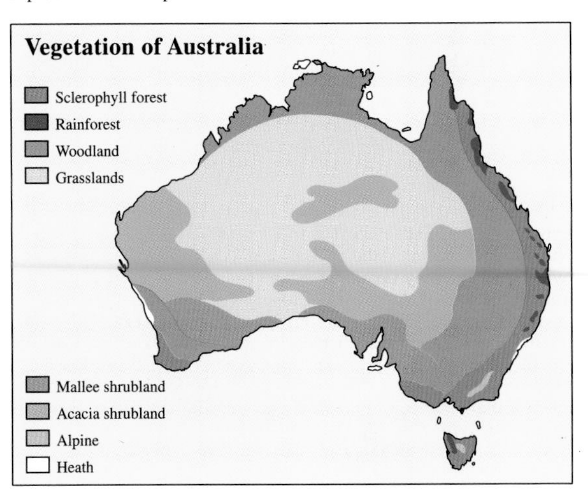

Rainforest

A rainforest is a closed forest dominated by closely spaced trees with dense crowns, forming an unbroken canopy of one or more layers of foliage. The closed canopy produces a perpetually shaded and humid interior capable of supporting a variety of epiphytes, orchids, treeferns and climbing plants. As the light on the forest floor is usually insufficient for many sun-loving plants, there is an absence of grasses and annuals.

In Australia, rainforests are confined on the mainland to the coastal strips and highlands of eastern Australia. Their total area is relatively small and their distribution is discontinuous, extending across northern Australia and far north Queensland to southern Victoria, rarely more than 160km inland. They also occur in large areas of south-western Tasmania. Rainforests grow on a wide range of soil types at altitudes ranging from sea level to 1200m.

Monsoon rainforests are found in small patches in tropical northern Australia from about Cooktown in Queensland, parts of the NT, to small pockets in the wetter areas in the Kimberley region, WA. Monsoon rainforests experience distinct wet and dry seasons and there is often a mixture of deciduous and evergreen trees. This vegetation type is structurally very variable, ranging from tall evergreen forests of 30m in height to deciduous thickets often only 2–3m in height on particularly dry sites. Robust woody lianes are often a prominent feature of monsoon rainforests and this type of forest may be referred to as vine thickets.

Tropical rainforests in Australia occur as a discontinuous strip in north-eastern Queensland between Proserpine and the Eungella Range behind Mackay and from around Townsville to the Cape York Peninsula. The tall straight trees often have prominent buttresses and large leaves with smooth margins. There is usually a deep and continuous canopy of two or three layers. The vegetation is dense and varied and a great variety of thick woody lianes, palms and strangler figs may be prominent. A number of tree species produce flowers and fruit directly from the trunks and branches (cauliflory). Epiphytes, such as ferns, orchids and aroids, are common on trunks and the larger branches of trees. The closed canopy produces a perpetually shaded interior and there is usually very little plant life on the forest floor.

Subtropical rainforests occur in patches from behind Mossman, Queensland, down to north-eastern NSW. There is a tall but uneven canopy of about 20 to 35m. The variety of plant species is approximately half of those found in tropical rainforests; epiphytes are present but less abundant, and fewer tree species have buttresses. There is also a greater tendency for a small number of tree species to show dominance in a subtropical rainforests and the plants have smaller leaves than tropical species. The mature canopy tends to let in more light and a multitude of smaller plants thrives on the forest floor.

Temperate rainforests occur in isolated pockets in the McPherson Range in southern Queensland, the New England Tablelands and Barrington Tops in NSW, and in parts of south-eastern Victoria and Tasmania. There is usually a simple one-layer canopy and often only one or two dominant trees. The leaves of shrubs and trees are smaller and they may have serrated edges. Lianes and other climbers rarely occur and orchids are few. Ferns and treeferns often dominate the ground layer and mosses and liverworts are common on the ground and on the trunks of trees.

Sclerophyll forest

This type of forest, also called open forest, is dominated by trees with sclerophyll leaves, mainly the eucalypts. The trees are usually of forest form with tall, straight trunks equal to, or greater in height than the depth of the canopy. The trees usually have flattish crowns and an open canopy cover of between 30 and 70 percent. Sclerophyll forests are common along the east coast of Australia, much of Tasmania and in the wetter part of south-west WA. The structure and composition of the understorey of open forest changes with rainfall, latitude and other environmental conditions. Two major types occur in Australia: dry sclerophyll and wet sclerophyll.

Dry sclerophyll forests are composed of medium-size trees that are usually less than 30m tall. Usually there are two or more species of eucalypts dominating the tree layer, while below there is a well-developed shrubby understorey. This type of forest can survive a fire; although the crowns of the trees may be burnt, they will regenerate from epicormic buds buried within the bark. Dry sclerophyll forests are characteristic of soils of low fertility and low or irregular rainfall.

Wet sclerophyll forests are tall open forests consisting of stands of eucalypts, often more than 60m in height. The broken canopy does not completely shut out light and the understorey may consist of small trees, shrubs and climbers. Treeferns are often prominent in moist places. The dominant eucalypts vary from place to place and include the giant Karri, *Eucalyptus diversicolor*, in south-west WA and the towering Mountain Ash, *E. regnans*, of Victoria and Tasmania. In sheltered gullies of warmer parts of eastern Australia, the Blue Gum, *E. saligna*, and Flooded Gum, *E. grandis*, may dominate. Wet sclerophyll forests often occupy an intermediate habitat between dry sclerophyll forests and rainforests, where eucalypts are absent.

Woodland

This vegetation type has a more open formation than sclerophyll forest. The trees are numerous but their trunks are shorter and the canopy cover is less than 30 percent. The dominant trees are chiefly eucalypts, occasionally she-oaks and species of *Callitris, Melaleuca* and *Acacia*. There are three main types of woodland in Australia: tropical woodland, which extends across northern Australia; temperate eucalypt woodland in the south-west and south-east; and semi-arid and arid woodlands farther inland. Open woodlands, where grasses are continuous and well developed, are known as savannah woodlands.

Shrublands

Shrublands are dominated by multi-stemmed shrubs or low trees, usually less than 8m tall. They occur in semi-arid regions where the annual rainfall is usually less than 300mm. Low shrubland is dominated by saltbush and bluebush shrubs up to 2m tall.

Mallee shrublands cover vast areas of southern Australia extending from western Victoria and south-western NSW across SA to WA. Trees are replaced by shrubby eucalypts with multiple stems arising from an underground rootstock known as a lignotuber.

Acacia shrublands are widespread plant communities of the arid parts of inland Australia, in which small trees or shrubs of the genus *Acacia* are dominant. The trees range from 2 to 10m in height and by far the most common is the Mulga, *Acacia aneura*, but other wattles of similar habit may be associated with it.

Grasslands

Grasslands occur mostly in arid and semi-arid zones. They consist of tussocks of perennial grasses that cover up to 70 percent of the ground surface. Trees may be sparse or absent.

Alpine

The alpine region is defined as an area above the tree line, characterised by low-growing shrubs, herbs and grasses forming a variety of communities such as herbfields, heath, feldmark, fens and bogs. The tree line is somewhat variable and usually occurs at around 1800 to 1830m elevation. This region is confined to the high mountain peaks of the Australian Alps in south-eastern Australia. Tasmania has many mountainous areas that are rich in alpine flora.

Heaths

In general, heaths are dominated by hard-leaved shrubs less than 2m in height. Trees forming a mallee-like habit are sometimes present. Heaths are usually coastal and adapted to salt-laden sea winds and extremely poor soils. Small pockets of specialised heaths are also found in alpine areas, tablelands and in semi-arid regions.

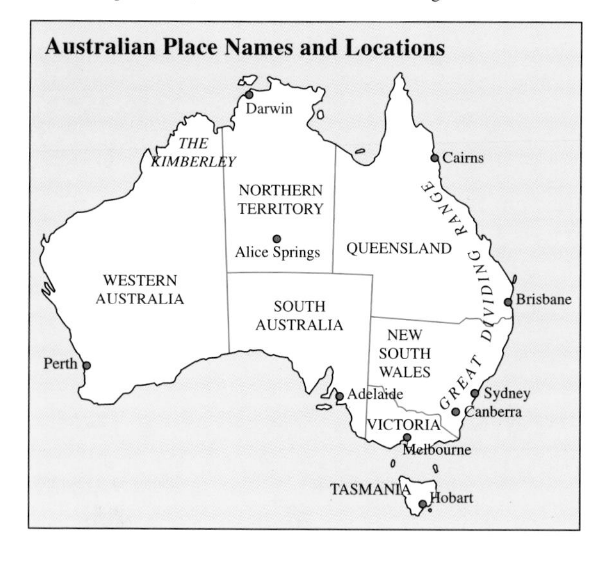

Key to symbols

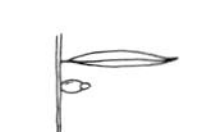

Brown Pine p. 10

Pines pp. 11–13

Cypress Pines
pp. 14–16

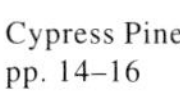

Sassafras pp. 17–18

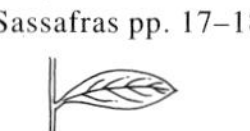

Bolly Gum p. 19

Kurrajongs pp. 20–23

Baobab p. 24

Kapok p. 25

Blueberry Ash
pp. 26–27

Hibiscus p. 28

Figs pp. 29–31

Stinging Trees p. 32

Euphorbias pp. 33–34

Beeches pp. 35–36

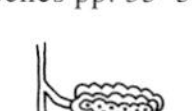

Casuarinas pp. 37–41

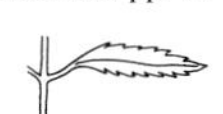

Coachwood Family
pp. 42–44

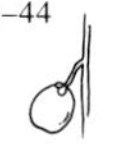

Davidson's Plum p. 45

Banksias, Grevilleas
etc. pp. 46–57

Eucalypt Family
pp. 58–99

Citrus Family
pp. 100–103

Cedar Family
pp. 104–106

Rosewood, Whitewood
and Tuckeroo p. 107–109

Acacia (Group 1)
pp. 110–112

Acacia (Group 2)
pp. 113–119

Acacia (Group 3)
pp. 120–121

Ironwood p. 122

Blackbean p. 123

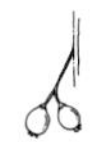

Sandalwood Family
pp. 124–126

Pittosporum
Family p. 127–129

Celery Wood,
Umbrella Tree
pp. 130–131

Palms pp. 132–135

Pandanus
pp. 136–137

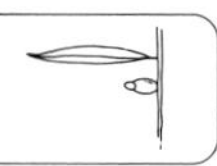

Brown Pine *Podocarpus elatus* 40m

Used extensively in carpentry and general interior work, the strong, fine-textured timber of the Brown Pine is among the most valuable of native softwoods. It is a medium-size tree, usually around 15m, found in coastal rainforests from near Nowra, NSW, to the Queensland border. It also occurs in isolated pockets in north-eastern Queensland,where it reaches its optimum height of 40m.

It has dark brown, fibrous and finely channelled **bark** that becomes scaly and shows a tendency to peel slightly on old trees. The crown is dense, but the foliage is unlike typical conifers. The dark green **leaves** are flat, oblong to linear, 5–14cm long and 1.5cm wide, and bear a prominent midvein. Attractive new growth in spring is a pale, yellowish-green. The tree is a successful ornamental species.

The **fruit** differs from other conifers and consists of swollen fleshy receptacles that rarely become woody. Male cones are in catkin-like clusters to 3cm long. Solitary female fruits (on separate trees) are bluish-black, covered with a waxy bloom and bear a globular seed about 1cm in diameter at the apex. The swollen receptacle is edible and has a plum-like colour and taste, which gives this tree its other common name of Plum Pine.

Queensland Kauri Pine *Agathis robusta* 50m

Noted as one of Australia's tallest conifers, this majestic timber-producing species can be seen as an emergent tree scattered through rainforests or as part of stands on the edge of rainforests. It has a restricted range and occurs in southern Queensland around Maryborough and Fraser Island and again in the Atherton district farther north.

This upright tree has a long straight trunk and smooth **bark** with grey, brown and orange tonings. The tree may have no branches for about two-thirds of its height, but the spreading crown is composed of numerous branches. The distinctive ovate **leaves** are dark green, flat and leathery with fine longitudinal veins; they may be up to 13cm long and 4cm wide.

The **fruit** is a typical pine cone with hundreds of overlapping scales. Male and female cones are borne on the same tree. Large female cones are ovoid, 9–15cm long and up to 10cm wide. Male cones are elongated and about 10cm long. The flat winged seeds are about 1cm long.

There are two other Australian species of *Agathis* and about 17 species occurring in New Zealand, south-west Pacific and Malaysia. The New Zealand Kauri Pine, *Agathis australis*, is a closely related species and, like the Queensland Kauri Pine, is highly valued for its beautiful, even-textured, pale brown timber used for joinery, cabinet work and other indoor uses. Kauri is a Maori name for *A. australis*.

Bunya Pine *Araucaria bidwillii* 45m

The large, symmetrical Bunya Pine, with its distinctive, dome-shaped crown, is a familiar tree of large parks and gardens throughout Australia. In nature, it has a restricted range, occurring as an emergent in rainforests in south-eastern Queensland near Gympie and in two isolated patches west of Cairns. Bunya Pine is famous for its large female cones, which provide highly nutritious seeds valued by Aborigines as a food source.

It is an upright tree with a straight, undivided trunk and persistent rough **bark** that is dark brown to almost black. The horizontal branches radiate from the stout trunk, generally with most of the foliage clustered at the tips. The ovate to lanceolate **leaves**, 5cm long and 1cm wide, are glossy dark green, sharply pointed and spirally arranged. The **fruit** is a hardy woody cone. Male and female cones are usually borne on the same tree. The cylindrical male cones, up to 10cm long, are produced at branch ends. Large pineapple-like female cones, to 30cm long and 20cm across, are produced high in the tree. Each of the woody cone scales bears a large, egg-shaped edible seed to about 3cm long.

Bunya Pine is a very valuable timber tree, but because of its slow growth is not as popular for forestry purposes as the related Hoop Pine.

Hoop Pine *Araucaria cunninghamii* 50m

This handsome, easily recognisable tree is a tall conifer with spirally arranged foliage clumped toward the ends of upswept branches. It occurs in coastal areas of northern NSW and Queensland, as well as Irian Jaya in Indonesia and Papua New Guinea. Hoop Pine is found in a variety of forest formations and is a conspicuous emergent in dry rainforest.

It has a straight trunk and its dark, rough **bark** has horizontal cracks that form hoops or circular bands. The crown is rather open. Small, linear, overlapping **leaves**, to 1.5cm long, have sharp incurved points. The almost rounded female **fruit** cone, up to 10cm across, has wedge-shaped woody scales. A single seed, about 1cm long, is embedded in each scale.

Widely planted for forestry purposes in Queensland and some overseas countries, the tree's valuable softwood is straight-grained and easily worked. It has been widely used for flooring, ceilings and other interior work.

Hoop Pine belongs to the ancient conifer family, *Araucariaceae*, of which the Bunya Pine and the widely planted Norfolk Island Pine, *A. excelsa*, are members. A newly described member of this family, the Wollemi Pine, *Wollemia nobilis*, was discovered in a remote part of a rugged national park near Sydney in 1994. There are fewer than 40 adult Wollemi pines and its exciting discovery has attracted worldwide interest.

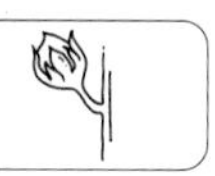

Coastal Cypress Pine *Callitris columellaris* 10m or more

Once included with the widespread White Cypress Pine, *C. glaucophylla*, until separated and named in 1984, Coastal Cypress Pine occupies different habitats and differs in appearance, habit, bark and foliage. It grows on sandy soils in subtropical coastal northern NSW and Queensland. A medium-size symmetrical tree, it has a single trunk, spreading branches and brown furrowed **bark**. The tiny dark green, scale-like adult **leaves** are 1–3mm long. Male **cones** are borne in clusters; smooth female cones, 2cm across, are solitary, with 6 scales. These separate almost to the cone base at maturity. The persistent 3-lobed central column, to 7mm long, is thick and angled.

White Cypress Pine *Callitris glaucophylla* 20 m

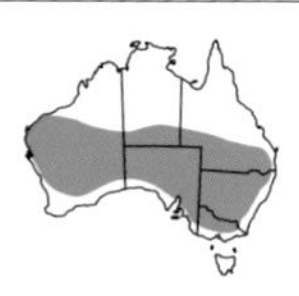

This widespread tree, formerly known as the inland form of *C. columellaris*, occurs in forest or woodland associations in all mainland States, mostly in inland districts. It is slender and erect with rough, greyish furrowed **bark.** Bluish-grey glaucous **leaves**, 1–3mm long, distinguish this species from *C. columellaris*. The solitary, slightly wrinkled female **cones**, 2cm across, are dark brown and consist of three large and three small alternating woody scales, all spreading widely at maturity. The 3-lobed central column is usually slender and less than 5mm long. The attractive termite-resistant timber is soft, durable and used widely for flooring and lining boards.

Stringybark Pine *Callitris macleayana* 20m

A medium-size tree found mostly bordering or in rainforests from the NSW north coast to south-eastern Queensland, the Stringybark Pine also occurs in a few localities in north-eastern Queensland. As its common name suggests, it has a fibrous, stringy, brown **bark**. The linear juvenile **leaves**, to 1.5cm long, are often produced over much of the crown on trees up to a considerable size. Adult leaves are tiny and scale-like, 3–6mm long. The pyramidal female **cones**, to 3cm across, have 6 or 8 scales of equal length, each with a small point near the tip.

Port Jackson Pine, Oyster Bay Pine *Callitris rhomboidea* 6m

A small green or greyish-green narrow shrub or small tree with disjunct occurrences in Queensland, NSW, Victoria, Tasmania and SA. It is the most common native conifer in bushland around Sydney and one of the most widely planted Australian conifers. It has coarse spreading branches with tiny keeled **leaves** to 3mm long. The new growth is slightly weeping.

The fruiting **cones**, to 2cm in diameter, are in tight persistent clusters. They are smooth at first, but become dry, wrinkled and dark brown with age. They have 6 thick scales, which are alternately large and small, each with a bold conical point.

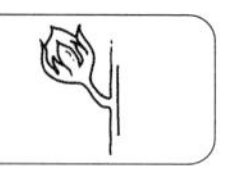

Rottnest Island Pine *Callitris preissii* 15–20 m

Although commonly known as Rottnest Island Pine, this species has a wide distribution across southern Australia and three subspecies are recognised. It is a branching, medium-size tree with dark grey fibrous **bark** and minute scale-like **leaves**, 2–4mm long, joined to the stem for most of their length. The female **cones**, carried on stout stalks, vary in the subspecies. In subspecies *preissii* (Rottnest Island Pine; see bottom photo) the cones are more or less rounded to 3cm across. Subspecies *murrayensis* (Murray Pine, top photo) has smooth ovoid cones to 3cm long with thick woody valves; while subspecies *verrucosa* (Mallee Pine) has smaller, slightly rounded cones, less than 2.5cm in diameter and covered in small prominent warts (tubercles). All cones have 3 large blunt scales and 3 small pointed ones.

The Murray Pine grows mainly along the Murray River Valley, NSW, Victoria and SA. The Mallee Pine is a small spreading tree, usually with several stems, often 2m high and rarely more than 5m. It favours semi-arid mallee areas in western NSW, Victoria and SA.

This tree was named in honour of German botanist Ludwig Preiss (1811–83), an early collector of plants in WA from 1828–42. All subspecies make excellent long-lived and drought-hardy trees for rural areas.

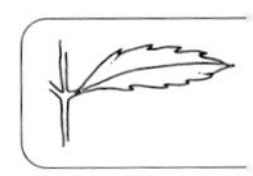

Southern Sassafras *Atherosperma moschatum* to 25m

The aromatic bark of this species has a distinctive sarsaparilla-like smell and taste, and was used by early settlers to make a tea substitute and tonic. Black Sassafras, widespread in cool temperate rainforests in Tasmania and Victoria, also occurs at altitudes above 1200m in scattered localities in NSW, often in association with Antarctic Beech, *Nothofagus moorei*. It can be found in shaded, protected situations.

It is a small to medium-size conical tree, usually 10–15m high, but occasionally attaining a height of 25m. The light grey-green and cream mottled **bark** is smooth. Young stems and flowers are usually hairy. Narrow ovate or lanceolate **leaves**, 3–10cm long and borne in opposite pairs, may be entire or sparsely toothed. Green and shining above and yellow-white beneath, they are strongly aromatic when crushed. The lightly scented cup-shaped cream or white **flowers**, 2cm across, face downwards and appear mainly in spring. They have 8 petal-like segments. The hairy **fruit**, with persistent styles, is enclosed in an enlarged cup-shaped receptacle that is 1cm across.

The easy-to-work timber is in demand for turnery and craftwork. It was once used for clothes pegs.

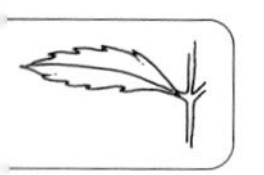

Sassafras *Doryphora sassafras* to 40 m

Sassafras is a common species found in both subtropical and warm temperate rainforests on the coast and ranges of NSW, extending from the Bega district to just inside the Queensland border, chiefly in the McPherson Ranges. At high elevations it is sometimes the dominant tree of the forest. The tree is aromatic in all parts and was named after the fragrant North American genus *Sassafras*. The aromatic bark of this species was valued by early settlers and infusions of it were used to purify blood and ease stomach ache.

This is a medium-size to tall straight tree with a relatively small compact crown. Young shoots and flowers are silky hairy. The grey **bark** is finely scaly. Coarsely toothed elliptic **leaves**, 3–10cm long and up to 4cm wide, are dark glossy green above and fragrant when crushed. White **flowers**, 3cm across, with prominent stamens, are borne in clusters, often in groups of 3, in the leaf axils in early spring. The dark brown hairy **fruit** is enclosed in an enlarged receptacle about 2cm long. The pale yellowish-grey, fine-textured timber is used for cabinet work, turnery and flooring.

Bolly Gum *Litsea reticulata* to 40m

Widely distributed in coastal districts from the Illawarra region of southern NSW to central Queensland, this species is a common component of warm-temperate and subtropical rainforests. Bolly Gum is a large tree with a dense canopy and grey mottled **bark** marked by shallow oval depressions and pinky-brown patches. The leathery narrow-oblong or oblong-elliptic, dark green **leaves**, 7–10cm long and 2–6cm wide, are rounded at the apex or slightly pointed and have prominent reticulated veins, hence the specific name, which means a network of lines. They are rather glossy and often very small oil glands can be seen. The margins are pale or translucent.

Small cream or greenish **flowers**, about 2.5mm long and borne in clusters of 6, are surrounded by deciduous bracts. They have a pleasant fragrance and appear mostly in early winter. The **fruit** is an oval black or purplish berry, to 13mm long. They ripen from summer to autumn and are most attractive to native birds. The pale brown timber is used for plywood, interior work and furniture making.

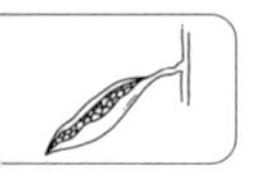

Flame Tree *Brachychiton acerifolius* 35m

Masses of fiery red flowers that can be seen at some distance in dense forests gave this tree its common name. It is an apt description for this flamboyant tree, which has the unusual habit of shedding most of its leaves around flowering time, highlighting its display. One of the most brilliant of Australia's flowering trees, the Flame Tree is cultivated in favourable climates throughout the world. It is naturally confined to coastal rainforests from the Illawarra region of NSW north to Cape York, Queensland.

It is a medium to tall spreading tree with a stout trunk and grey fissured **bark**. The inner bark, composed of lace-like fibres, was used by Aborigines for making cord and netting. The shiny, bright green **leaves**, on long slender stalks, are entire or variably lobed and maple-like, up to 20cm long and 25cm wide. The name *acerifolius* is from the maple genus *Acer*.

Masses of bright red, waxy, bell-shaped **flowers**, 1.5cm wide, are borne in large hanging panicles from November to February. The **fruit** is a dark brown, boat-shaped leathery pod, to 15cm long, which splits open to reveal rows of seeds surrounded by outer seed-coats that bear numerous fine irritating hairs.

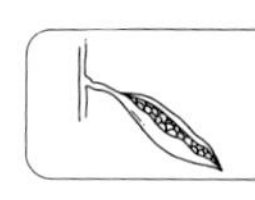

Lacebark Tree *Brachychiton discolor* to 30 m

Found in dry rainforests and occasionally sub-tropical rainforests north from the Hunter Valley, NSW, to around Maryborough, Queensland, this species has large pink flowers, but grows into a tall deciduous tree. It has grey, prominently fissured **bark** on its slightly swollen trunk. The large, wide, adult **leaves**, up to 20cm long, have five rather shallow lobes and a whitish, woolly underside. When juvenile, these are distinctly 5 to 7 lobed and softly hairy all over. In summer, deep pink, bell-shaped **flowers** appear, up to 5cm across, and velvety on the outside. The **fruit** is a brown woody, boat-shaped follicle, 10–15cm long.

Red-flowered Kurrajong *Brachychiton megaphyllus* 3–6m

Clusters of up to 30 orange-red flowers make this small, straggly deciduous tree conspicuous during the dry season (June-October) in tropical woodlands of the NT. It has rough, dark grey furrowed **bark** and large, rough, wide **leaves**, up to 30cm long and with 3–5 lobes, appear at the beginning of the wet season. Showy bell-shaped, 5-petalled **flowers** are produced on very short stems on the old wood. The **fruit**, a yellowish, oblong woody follicle, is covered in soft bristly hairs. When ripe, it splits open along one side to reveal many yellow seeds, which are eaten raw or cooked by the Aborigines. This Kurrajong is Darwin's floral emblem.

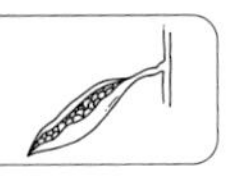

Kurrajong *Brachychiton populneus* 15m

Kurrajong has a wide distribution in eastern Australia, occurring in coastal areas as well as on the western side of the Great Dividing Range from inland and southern Queensland, NSW and north-eastern Victoria.

It is a shapely, densely crowned tree with grey fissured **bark** and a trunk that is almost barrel-like near the base. The juvenile **leaves** often have 3–5 lobes and differ markedly from the mature ovate leaves, which are mostly entire, 7cm long and 3cm wide, or sometimes 3-lobed. Attractive creamy white, bell-shaped **flowers** with red throats, about 2cm long, are borne profusely throughout spring. The **fruit** is a thick-walled, boat-shaped follicle, 2–7cm long, that tapers to a hooked beak. When ripe, the fruit splits open to reveal numerous seeds. These were collected and eaten either raw or roasted by Aborigines. When roasted and ground, they make an agreeable coffee-like beverage.

A close relative of the Kurrajong is the Queensland Bottle Tree, *B. rupestris*, found in drier parts of central Queensland. It is distinguished by its often prominently bottle-shaped trunk with dark furrowed bark. The smaller Desert Kurrajong, *B. gregorii*, to 8m high, is found scattered throughout the desert areas of NT, SA and WA. All three species are important fodder trees in times of drought and, when tapped, the roots rapidly yield large quantities of water.

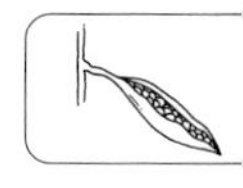

Brown Kurrajong *Commersonia bartramia* 5–12m

Because of its ability to colonise cleared land rapidly, this is an important small tree in rainforest regeneration. It is most often seen at the edge of rainforests and in clearings and is especially noticeable in summer when masses of small creamy white flowers weigh down the almost horizontal branches. It is sometimes called the Scrub Christmas Tree. It occurs on the coast and ranges north from the Bellinger River, NSW, to Cape York Peninsula, Queensland, extending to the Pacific Islands and South-East Asia.

The branches are widely spreading and tend to give the tree a layered appearance. Young branches and twigs are covered with numerous white corky spots. The ovate to broad-ovate **leaves**, 6–15cm long and 4–10cm wide, are shallowly toothed or sometimes almost entire. The undersides are silvery yellow with minute hairs. The small but prolific **flowers** have conspicuous stamens and are massed in erect clusters on top of the branches. The **fruit** is a rounded capsule, about 2.5cm wide, covered with numerous soft bristles.

The bark fibre of brown kurrajong was used by Aborigines for making nets. This ornamental, fast-growing species is popular for growing in warm temperate and subtropical gardens.

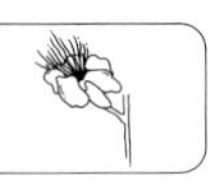

Baobab, Boab *Adansonia gregorii* 15m

This distinctive and often grotesquely shaped deciduous tree is common in the Kimberley region of WA, extending into the Victoria River district in the NT. It grows in open woodlands, on sandy plains and rocky ridges, often in pure stands.

The Baobab is noted for its huge barrel-shaped trunk, sometimes reaching 20m in circumference. The smooth **bark** is light brown or pinkish-grey. The spreading branches are leafless for much of the dry season. At the start of the rains it bears large palmate **leaves** with 5–9 pointed elliptic leaflets about 13cm long and 3.5cm wide. The lower surface of the leaves is covered by short soft hairs. During the early wet season, large creamy white **flowers** with 5 fleshy petals and numerous protruding stamens are produced in the leaf axils. The **fruit** is a gourd-like woody capsule, 20cm long and 10cm or more across, covered with short yellowish-green hairs. It contains many kidney-shaped seeds embedded in whitish pith — both seeds and pith are relished by Aborigines. The fruit pulp is eaten for gastric disturbances.

Depressions formed at the base of larger branches collect and store water, which is used by Aborigines in times of drought. Fibre obtained from beaten roots is used to make string and nets. The wood is extremely soft and spongy and cattle have been known to eat most parts of a fallen tree.

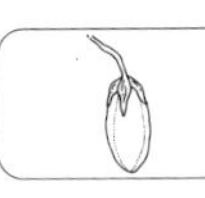

Kapok Tree *Cochlospermum fraseri* to 6m

This slender shrub or small spindly tree is a common understorey plant of tropical open forests and woodlands throughout the Kimberley region of WA and the Top End of the NT. It is deciduous and has smooth green **bark** and dark green **leaves**, 8cm long and 12cm wide with 5–7 variable lobes. The margins are shallowly toothed. New leaves are produced at the beginning at the wet season. During the dry season, showy bright yellow **flowers** are produced while the plant is still leafless. These are 6–8cm across and have 5 sepals covered in soft hair, 5 petals and numerous stamens.

The **fruit** is a smooth oblong capsule, 8–12cm long. It is initially green, but ages to brown in early summer when it opens to release many kidney-shaped seeds that are surrounded by a mass of cottony fibres. Aborigines use the woolly material of the fruit to decorate their bodies for ceremonies. The roots of young plants are eaten and the green unripe fruit is broken and used for localised skin infections. The young stems are used as fire-sticks, while the green bark makes a fairly strong string and is sometimes used as a paint brush.

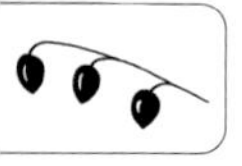

Blue Quandong *Elaeocarpus grandis* 35m

This majestic tree is found mostly beside running water in rainforests along the eastern coast, from the Nambucca River, NSW, to northern Queensland. Its canopy is open and airy; the trunk is prominently buttressed at the base; the smooth **bark** is whitish. The rather stiff, oblong-elliptic **leaves**, 8–19cm long and 1–4cm wide, are finely toothed. Old leaves turn bright red before dropping. In autumn, pendent, greenish-white fringed **flowers** are carried on one side of the raceme, followed in winter and spring by bright blue oval **fruit**, about 3cm in diameter, with a thin layer of edible flesh around the stone. These are attractive to fruit pigeons and are also eaten by Aborigines. This species is sometimes included with *E. augustifolius*.

Blueberry Ash *Elaeocarpus reticulatus* 5–15m

This is a widespread and common small tree of sheltered forests and rain forests of eastern Australia extending from Fraser Island, Queensland, south to Tasmania. It has brown **bark** with vertical fissures. Oblong-elliptic **leaves**, 5–13cm long and 1–3cm wide, have regularly toothed margins and prominent reticulate net veins, which separates this species from *E. grandis*. The hanging bell-shaped white or sometimes pink **flowers** resemble lily-of-the-valley, but with dainty fringes. They have an aniseed-like perfume. The **fruit**, a blue ovoid drupe about 12mm long, ripens in winter and is sought after by bower-birds and currawongs.

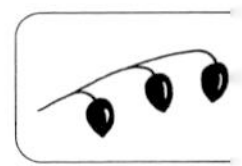

Yellow Carabeen *Sloanea woollsii* Up to 55m

In subtropical rainforests, few tree species have buttresses and these aboveground thickened root systems are usually associated with only one or two species of figs. A notable exception in rainforests of NSW is Yellow Carabeen, which is easily recognised by its very large plank-like buttresses, sometimes extending 4m up the trunk. It occurs north from the Barrington Tops to south-eastern Queensland.

Yellow Carabeen is a tall tree with a greyish-brown **bark** that has small raised corky spots in vertical lines. It has bluntly toothed, elliptic-lanceolate **leaves**, 7–19cm long and 2–6cm wide, with a distinct basal joint. Tufts of brown hairs are present in the axils of midrib and lateral veins. New foliage is yellowish-green. The petalless **flowers** are borne in axillary racemes in late spring and have 4–5 velvety sepals and about 24 bristle-like stamens. The yellowish-brown **fruit** is a woody 2-valved capsule, about 2cm long, coated with slender prickles.

There are only 5 species of *Sloanea* in Australia. Yellow Carabeen differs from the closely related Maiden's Blush, *S. australis*, in that the flowers lack petals and brown hairs are present in the vein-angles.

Norfolk Island Hibiscus *Lagunaria patersonii* 12m

Although mainly associated with Norfolk Island and Lord Howe Island, this handsome tree is widely cultivated in temperate areas and, because of its ability to withstand strong, salt-laden winds, is often a favoured street tree in coastal districts.

This is a monotypic genus named after Andrea Laguna, a 16th-century Spanish botanist. It has a densely foliaged pyramidal outline, with branches almost to ground level and a rough, dark grey **bark**. The thick, ovate, olive green **leaves,** 6–10cm long and up to 6cm wide, have a whitish underside. The pale pink to mauve, open bell-shaped **flowers**, 6cm across, have 5 recurved petals and are produced mainly through summer and autumn. These appear in great numbers and attract many varieties of nectar-feeding birds. The flowers resemble the small, single-flowered hibiscus, to which the tree is related. The **fruit**, an ovoid capsule 2–3cm long, contains dark red seeds. The capsules contain numerous fine splinter-like hairs that penetrate the skin and cause irritation, so care should be taken when removing the seeds.

Propagation is from seed, which germinates easily. When grown as a screen in seaside gardens, this tree will give excellent protection for plants that are less salt-tolerant.

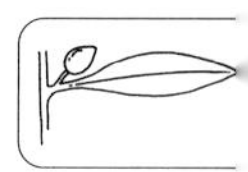

Moreton Bay Fig *Ficus macrophylla* up to 50m

Familiar from parks and large gardens, this very large strangling tree is found in rainforests from southern Queensland to southern NSW, eventually forming a massive trunk with buttressed and latticed roots. It has a thick spreading crown and smooth greyish-brown **bark**. The large, ovate to oblong glossy green **leaves**, 10–25cm long and 7–10cm wide, are covered with reddish-brown hairs on the undersides. Large rose-coloured leaf-sheaths (stipules), to 17cm long, enclosing the buds appear at branch ends. The globular fleshy **fruit**, about 2.5cm across, are borne singly or in pairs on thick stalks. Orange at first, they turn purple when ripe and are covered in white dots. They are relished by birds and flying foxes.

Small-leaved Fig *Ficus obliqua* 15–35m

This medium-size to very large tree is widespread in subtropical and littoral rainforests from southern NSW to north-eastern Queensland. It has smooth greyish-brown **bark** and its massive, buttressed trunk is often laced with strangling roots. The relatively small elliptic **leaves**, 3–10cm long and 2–6cm wide, have distinct lateral veins that are more prominent on the undersides. Globular orange **fruit** with dark red spots, 1cm across, are borne in pairs in the leaf axils.

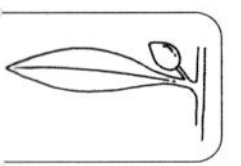

Cluster Fig *Ficus racemosa* 12–20m

The Cluster Fig has a wide distribution in northern Australia, in Queensland, the NT and the Kimberley region of WA. Its range includes India through South-East Asia to New Guinea. It grows along creeks and rivers in monsoon forests and on the edges of coastal plains. The trunk is often buttressed at the base; the **bark** is smooth and pale grey. The ovate-elliptic **leaves,** 7–18cm long and 3–8cm wide, are pointed at the end and have prominent, well-spaced veins. The **fruit**, to 4cm across, is produced in large clusters from the trunk and main branches. Although edible, the figs are not as sweet as other varieties. They are yellow to orange-red when ripe, usually at the end of the dry season.

Rock Fig *Ficus platypoda* to 6m

This spreading shrub or small tree is widespread across tropical northern Australia from the Kimberley, WA, in Central Australia and Queensland as far south as Brisbane. It grows on coastal cliffs, among sandstone, on rocks and in crevices on rocky outcrops, usually in association with a watercourse or water hole. It has smooth pale grey **bark**. Aerial roots and exposed spreading roots may be present. The leathery, broadly ovate, prominently veined **leaves,** 6–9cm long and to 5cm across, are borne on broad stalks. The **fruit**, to 1.5cm across, usually occurs in pairs in the leaf axils. Yellow at first, the figs turn orange, red and purple when ripe, usually early in the dry season (autumn). Juicy and sweet, they are eaten by Aborigines.

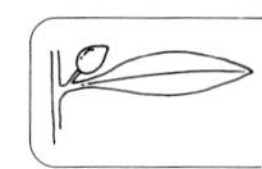

Port Jackson Fig *Ficus rubiginosa* 10–25m

Widespread and common, this small to medium tree extends along the coast from near Bega, NSW, to south-eastern Queensland, inhabiting the drier margins of rainforests. It also occurs slightly inland on the northern tablelands and western slopes of NSW, where it is often found in quite dry situations on rocky slopes and in open forests. It has a short, irregularly buttressed trunk with pale yellowish-brown **bark**. The leathery ovate-elliptic **leaves**, 7–10cm long and 5–6cm wide, are rust-coloured with hairy undersides in most forms. Mature **fruit**, 2cm across, are borne on short stalks, mostly in pairs. Yellow at first, they ripen to deep red; they usually have warts and a short nipple at the tip.

Strangling Fig *Ficus watkinsiana* 25–35m

This large tree is a common strangling fig in rainforests from northern NSW to north-eastern Queensland. It has grey **bark** and a massive buttressed trunk, often laced with strangling roots. It has dark, reddish-purple stipules to 6cm long and shiny dark green elliptic **leaves**, 10–25cm long and 10–15cm wide, with a paler underside. The mature dark purple ovoid **fruit**, each about 4cm long, has pale spots and a prominent nipple. These are often found on the forest floor and help to separate this species from other fig trees.

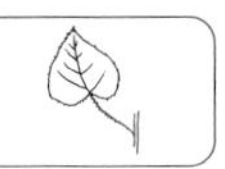

Giant Stinging Tree *Dendrocnide excelsa* 10–40m

Among the most troublesome plants of eastern Australia's rainforests are 4 species of stinging trees. When brushed against, very fine stinging hairs on leaves, leaf stalks and young branches penetrate the skin, introducing an irritating fluid and causing intense pain. Young trees appear in disturbed areas where an opening in the canopy lets in extra light. In denser rainforest situations the giant stinging tree sometimes persists as a canopy tree and can reach huge proportions with a massive buttressed trunk. It has pale brown scaly **bark** and large broad-ovate **leaves**, 10–25cm long and 7–20cm wide, with a heart-shaped base and downy underside. The leaves, branchlets and leaf stalks are covered with stinging hairs. Short axillary clusters of greenish-white male or female **flowers** appear on separate trees. The **fruit** is a cluster of small cream or mauve fleshy stalks with a small dry nut on each.

Gympie Stinger *Dendrocnide moroides* 4m

The most notorious of the stinging trees, the Gympie Stinger is a tall shrub or small tree, which is common on rainforest margins of northern NSW to northern Queensland. It is said to inflict the most painful sting. This species differs from *D. excelsa* by the broad-ovate **leaves**, which have the stalk inserted on the blade instead at the edge (peltate). The **fruit** is a mass of pink to purplish fleshy stalks with a small dry nut on each.

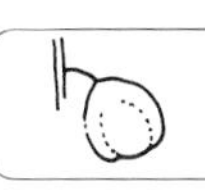

Bleeding Heart *Omalanthus nutans* to 6m

Large openings in the rainforest canopy can be made by a cyclone or some other disaster, and this allows extra sunlight to reach the forest floor. Herbaceous plants can seed quickly and spread rapidly. These are gradually replaced by fast-growing pioneer shrubs and small trees that can withstand exposure while protecting the slower growing primary tree saplings. These, in turn, may take decades to become established. Bleeding Heart, formerly *O. populifolius*, is a common regeneration shrub or small tree of the rainforest edge and clearings and is widespread along the coast of NSW to central Queensland.

Young stems and leaf stalks of this soft-wooded plant exude a whitish sap. **Leaves** are broad-ovate, 3–15cm long and 3–12cm wide. They have a satiny sheen and resemble those of the exotic poplar tree. Indeed, it is sometimes called the Native Poplar. The old leaves turn a brilliant red just before they are shed and most plants have a few coloured leaves scattered through the foliage. In spring, tiny green **flowers** are borne in terminal racemes about 10cm long. The **fruit**, a pale-green, 2-lobed capsule about 1cm across, ripens in summer and is a favoured food of the brown pigeon.

With its fast growth and colourful foliage, Bleeding Heart is becoming popular as a garden subject and indoor plant.

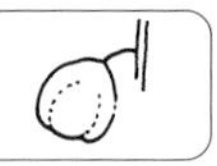

Quinine Tree *Petalstigma pubescens* 6m

This small tree, which shows some variation in size, has a very wide distribution from the Kimberley region of WA, across northern Australia, throughout most of Queensland and into north-eastern NSW. It is also recorded for several offshore islands and for New Guinea. This is a fairly common understorey tree in open forests and woodlands.

The Quinine Tree has a rounded spreading canopy and a finely fissured dark grey to black **bark**. The ovate to somewhat circular **leaves**, 2–6cm long and up to 3cm wide, are shiny dark green above with a finely pubescent whitish underside and a rounded to slightly pointed tip. Tiny white to creamy yellow male or female **flowers** are borne on separate plants, the male flowers in clusters of 3 or 4, while female flowers are solitary. The globular **fruit**, to 2cm in diameter, is bright orange and somewhat succulent at first. When mature, it splits into several segments and sheds several dark brown seeds.

The fruits are unpalatable with a very bitter taste, but are used in traditional Aboriginal medicine. The peelings are soaked in water, which is drunk by women to prevent unwanted pregnancy. A decoction of the fruit is also applied to aching teeth.

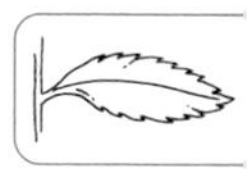

Myrtle Beech *Nothofagus cunninghamii* 30–40m

Of a world total of 35 species of *Nothofagus*, 3 are endemic to Australia. The rest occur only in the southern hemisphere in small pockets of New Zealand, New Guinea, New Caledonia and South America. Fossil evidence shows that *Nothofagus* once formed extensive forests on the supercontinent Gondwana. The 3 Australian species are considered remnants from previously extensive rainforests on the eastern and southern coasts during the early part of the Miocene, about 20 million years ago.

Myrtle Beech is the dominant canopy tree of the beech forests of eastern and central Victoria and western Tasmania where, in moist sheltered conditions, it grows up to 40m with many spreading branches. It has slight buttresses or basal burls and dark brown scaly **bark**. The small, toothed ovate **leaves**, 1–2cm long and 1–1.5cm wide, are marked by translucent glands and form a dense, lacy canopy. In spring, new leaves are a conspicuous bronzy pink. The **flowers** are small and inconspicuous. Male flowers are usually solitary with a short stalk and numerous stamens. Female flowers with protruding styles are borne in trios above the male near branch ends. They appear in spring and the wind distributes the pollen. The **fruits** are 3 small dry nuts enclosed in a bristly, 4-valved body.

Most beech forests are now protected in national parks and represent some excellent examples of old-growth forests.

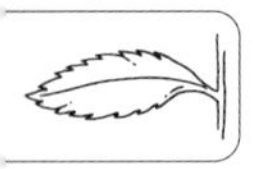

Antarctic Beech *Nothofagus moorei* 30–40m

Early bushmen also called this tree Negrohead Beech or Niggerhead Beech, an allusion to the dense, dark green crown. It has been variously estimated that some Antarctic Beech trees could be as old as 3000 years. In cool temperate rainforests of the McPherson Ranges, Queensland, and the Barrington Tops, NSW, some wonderful ancient specimens with huge gnarled, moss-covered trunks can be seen. The tree is confined to higher elevations and at about 1500m it forms fairly pure stands.

The gnarled trunks often have multiple stems consisting of suckers of various ages. The dark brown **bark** is rough and flaky. Glossy, dark green ovate **leaves**, 3–10cm long, have evenly toothed margins and several lateral veins. Separate male and female **flowers** are borne on the same tree. Small cup-shaped male flowers with numerous stamens are formed below the female flowers, in groups of 3, surrounded by numerous small bracts. The **fruits** are 3 small dry nuts enclosed in a bristly, 4-valved body.

The pinky-brown timber is of commercial value, but fortunately most of the trees are now included in national parks.

If you look carefully, you may spot the Beech Orchid, *Dendrobium falcorostrum*, which grows as an epiphyte high up in the branches. The flowers of this orchid are white with purple marking and have a delightful fragrance.

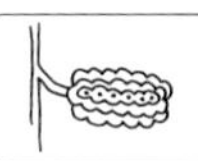

Belah *Casuarina cristata* 10–20m

A familiar sight in inland NSW and central Queensland, Belah is a medium-size tree with somewhat drooping foliage and hard, dark brown scaly **bark**. Thin branchlets of all casuarinas have fine ridges and joints at intervals, around which are tiny teeth-like scales in place of **leaves**. The number and spacing of these teeth help to separate species. They can be seen with the naked eye, but are better examined with a hand lens. Belah has grey-green branchlets with whorls of 8–12 **leaf-teeth**. Male and female **flowers** are on separate trees, male flowers in slender, terminal spikes 2.5cm long, female flowers clustered in tiny heads. The globular fruiting **cone**, about 2.5cm long, has protruding triangular valves.

River Oak *Casuarina cunninghamiana* 15–35m

In the past, its reddish-brown, close-grained timber was used for bullock-yokes and shingles, but the River Oak is now totally protected in NSW because of the way it stabilises river banks and prevents erosion. Also known as River Sheoak, this stately tree is common along the banks of freshwater rivers in eastern Australia from southern NSW to northern Queensland. The **bark** is hard, dark grey and finely fissured; the slender branchlets have 8–10 pointed **leaf-teeth** in whorls at each joint. Male and female **flowers** are on separate trees, male flowers in dense terminal spikes, about 1.5cm long, female flowers in a compact head. Small fruiting **cones** are about 1cm long and 8mm across.

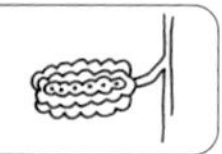

Coastal Sheoak *Casuarina equisetifolia* 6–12m

Its graceful weeping branches, tipped with silvery tufts, make this one of the most beautiful of all sheoaks. This usually small tree extends along the coast from near Port Macquarie, NSW, to northern Queensland and the Top End of the NT. Its range extends to islands north of Australia, including New Caledonia, New Guinea and Malaysia. Coastal Sheoak is widely planted in eastern Australia for sand stabilisation and reclamation of land after beach-mining activities. Its ability to withstand the full blast of salt-laden winds and its attractive appearance also make it a popular street tree in warmer seaside towns.

The smooth grey-brown **bark** has been used in tanning and for the extraction of dyes. The ribbed, grey-green branchlets have 6–8 finely pointed **leaf-teeth** in whorls around the stem at each joint. Male and female **flowers** are on the same plant, with the male flowers in short spikes, to 1.5cm long, at the ends of the branchlets. The small, cylindrical hairy **cone**, 1–2cm long, has a flattened top and sharply pointed valves.

The dark-coloured timber is used for ornamental turnery and for parquetry. Being relatively smokeless and producing good heat, the wood makes an excellent fuel. In traditional Aboriginal medicine, the inner bark and young sapwood is infused in water and used as a mouthwash for toothache and sore gums.

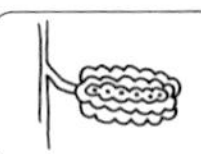

Desert Oak *Allocasuarina decaisneana* 5–18m

Occurring in arid areas of central Australia, this tall, open-branching tree offers welcome shade and beauty in hot desert regions. Young plants are erect, but as the tree matures it develops a graceful drooping habit. Its dark brown, deeply furrowed **bark** is fire resistant. Whorls of 4 pointed **leaf-teeth** occur on rather short deciduous branchlets. The hairy seed **cones** produced on the female trees are the largest in the genus, up to 7cm long and 2.5cm wide, and have sunken valves. Hollow tree trunks of the desert oak store supplies of water valued by Aborigines travelling through dry country.

Western Australian Sheoak *Allocasuarina fraseriana* to 15m

This medium-size, handsome tree is common in woodlands and open forests, often in association with Jarrah, *Eucalyptus marginata*, in the south-west of WA. It has fine-textured, slightly flaky grey **bark**. The crown is moderately large with deep green, often pendulous branchlets. **Leaf-teeth** in whorls of 6–8 have fringed margins. Male and female **flowers** are on separate trees. The golden-brown male flowers are borne profusely in slender terminal spikes, to 10cm long, in early spring. Clusters of female flowers eventually form a somewhat globular fruiting **cone**, 2–4cm long, with a wrinkled surface.

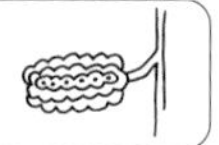

Black Sheoak *Allocasuarina littoralis* 5–15m

This common erect tree of open forests and woodlands from the coast to the mountains extends from southern Tasmania to Cape York Peninsula, Queensland. The common name alludes to the dark or blackish appearance of the tree at certain times of the year. It has dark grey fissured **bark** and slender branchlets. **Leaf-teeth** are in whorls of 6–8. Male and female **flowers** most often occur on separate trees. In autumn, rusty brown male flowers appear in terminal spikes to 5cm long. Small red female flowers, in ovoid heads, eventually form a cylindrical woody fruiting **cone** 1–3cm long, with blunt, rounded valves.

Bulloak *Allocasuarina luehmannii* 5–15m

This medium-size tree of semi-arid regions has a rather stiff, upright appearance. It has a wide distribution and is commonly seen in open woodlands in Queensland, NSW, Victoria and SA. The **bark** is dark, deeply fissured and coarse, and the slightly waxy branchlets have **leaf-teeth** in whorls of 10–14. Male and female **flowers** are on separate trees, the yellowish male flowers in spikes 2.5cm long, borne toward the ends of branchlets. Female flowers, in short compact heads, eventually form a flattened **cone**, about 1.5cm in diameter with only 2–3 rows of rounded valves. Flowering is mainly in summer.

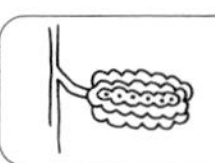

Forest Oak *Allocasuarina torulosa* 12–20m

The graceful Forest Oak is found along the eastern Australian coast and ranges from Nowra, NSW, to Cairns, northern Queensland. It occupies a variety of habitats, including rainforest margins, but is common in open forests. In certain conditions, the slender drooping branchlets are a most attractive coppery colour. The deeply furrowed light brown **bark** has a corky appearance. **Leaf-teeth** are in whorls of 4 or 5. Male and female **flowers** are on separate trees, male spikes, about 3cm long, at the ends of deciduous branchlets. The almost rounded warty **cones**, to 3cm long and 2.5cm wide, have a flattened top and broad rounded valves.

Drooping Sheoak *Allocasuarina verticillata* 4–10cm

This small bushy tree, formerly known as *Casuarina stricta*, is widely distributed in southern Australia through NSW, Victoria, SA and Tasmania. It occurs in a variety of habitats, from dry stony ridges and rocky outcrops inland, mountain slopes, grassy woodland or open forests, to exposed coastal cliffs. It has a rounded weeping crown and the **bark** is dark grey, furrowed and rough.

Whorls of 9–13 **leaf-teeth** appear on slender dark green, prominently ridged branchlets. Male and female **flowers** are on separate trees. Golden-brown male flower spikes are up to 12cm long; female flowers in short heads form cylindrical **cones**, to 4cm long, with numerous rows of pointed valves.

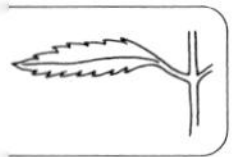

Soft Corkwood *Caldcluvia paniculosa* 10–20m

The very corky, pale fawn-coloured **bark** of this species was used at one time as a cork substitute, which included stoppers for bottles. This medium-size tree is common on the edge of rainforest communities as a pioneer species, and sometimes persists as a canopy tree. Its range is from the Hawkesbury River, NSW, to near Mackay in Queensland. Some beautiful mature specimens can be seen at Barrington Tops, NSW.

The Soft Corkwood is a straight tree with a slender, unbuttressed trunk and characteristic slightly fissured corky bark. The pinnate **leaves** are composed of 3–7 leaflets, each leaflet 5–15cm long and 2–4cm wide, ovate, dark green and with a toothed margin. Flushes of new growth are an attractive bronzy pink. In spring, tiny white **flowers** are borne in panicles 10–15cm long. The **fruit**, a tiny reddish ovoid capsule, is produced in large numbers in a terminal panicle.

Because of its showy flowers and decorative, coloured foliage, this species is popular in cultivation, especially in tropical and warm temperate areas,where it can grow quite quickly. It was formerly known as *Ackama paniculata*.

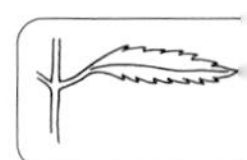

Black Wattle *Callicoma serratifolia* 3–10m

This tree is not really a wattle, a common name now generally applied to the genus Acacia. The common name comes from early colonial days in Sydney when settlers used its long, pliable branches to make wattle-and-daub huts. *Callicoma* is a monotypic genus endemic to Australia. This species is common on the margins of rainforests and frequently grows along creeks and moist gullies, often forming thickets. It is widespread along the coast and ranges from the Blue Mountains, NSW, to southern Queensland.

It is a multi-branched shrub or small tree with a moderately dense canopy and a smooth grey **bark** that is often spattered with colourful patches of lichen. The shiny, dark green elliptic **leaves**, 3–12cm long and 3–5cm wide, have a felted, silvery underside and coarsely toothed margins. Fluffy, creamy white **flower** heads, 2cm across, which resemble those of wattle, are produced during spring. The **fruits** are small capsules clustered in rounded heads. Popular in cultivation, this highly decorative plant develops into a large compact shrub that starts to produce its showy flowers when quite young.

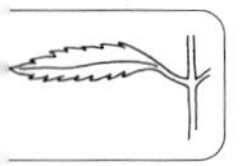

Coachwood *Ceratopetalum apetalum* 10–25m

Coachwood occurs as scattered trees or in almost pure stands and may be the dominant tree species in warm temperate rainforests from the McPherson Ranges on the Queensland border south to Batemans Bay, NSW. The largest intact area of Coachwood rainforest is in the Washpool National Park, northern NSW.

It is an important timber tree with a straight trunk. Its smooth, whitish **bark** with conspicuous circular rings is often blotched with grey and white lichens. Dark green **leaves** are oblong-elliptic, 6–12cm long and 2.5cm wide, with shallowly toothed margins. Greenish-white **flowers** with 5 petal-like sepals are borne in dense clusters from spring to summer. Like the closely related and commonly cultivated Christmas Bush, *C. gummiferum*, the sepals enlarge and become red when in fruit. The leaves of the Christmas Bush have 3 leaflets, which helps to separate the 2 species. The **fruit** is a dry one-seeded nut surrounded by the enlarged sepals.

Coachwood is frequently cut as a timber tree, its fine-textured, pinkish-brown timber being used for cabinet-making and interior work. During World War II, it was logged extensively to provide wood for rifle butts and veneers for the bodies of Mosquito fighter-bombers. The common name refers to the former use of the timber for coach-building.

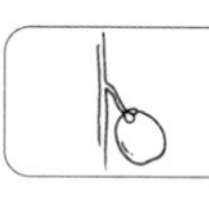

Davidson's Plum *Davidsonia pruriens* 6–10m

This small, slender rainforest tree occurs in two separate localities, one in north-eastern Queensland and another in the Brunswick and Tweed River areas, NSW. It is quite rare in the wild, but is a popular garden subject, famous for its large purple fruit, which resembles plums and was used by early settlers for jam-making.

The brown **bark** is corky and scaly. The large pinnate **leaves** have a winged axis and 7–17 oblong leaflets, each up to 25cm long and 9cm wide, with sharply toothed margins. Stems and leaves have pale brown hairs that can cause irritation if handled — *pruriens* means itching. In spring, small, brownish **flowers** are borne in narrow panicles up to 10cm long. These are followed by blue-black, plum-like oval **fruits** with soft red juicy flesh and relatively small flattened seeds with fibrous coats. The fruit matures late in summer and has a rather sour taste, but when stewed with sugar makes a delicious jam or jelly.

This species is mainly grown in warm temperate and tropical regions for its ornamental foliage, both in the garden and as an indoor or container plant. Flushes of new growth are colourful and highly decorative. It needs protection from extreme frost and heat and will bear fruit in 3 or 4 years.

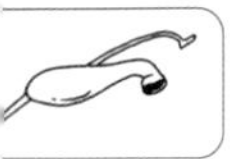

Satin Oak *Alloxylon flammeum* 20–30m

Alloxylon is a relatively new (1991) genus of 3 species endemic in eastern Australia and one in New Guinea and the Aru Islands. Australian species were formerly included in *Oreocallis*, which now comprises two species found only in Peru and Equador. *A. flammeum*, which has been for many years confused with *Oreocallis wickhamii*, is described as a new species.

It is a large, tall tree with a spreading crown, found in dense rainforests at high elevations on the Atherton Tablelands in north-eastern Queensland. Vast areas of its habitat have been cleared for land use and the tree is now quite rare in the wild. The grey **bark** is lightly fissured with raised corky spots. Thin-textured, narrow-elliptical adult **leaves**, 8–25cm long and 2–4.5cm wide, have entire margins and are light to dark green. Immature leaves are shallowly to deeply lobed. Bright orange-red racemes of 10–52 individual flowers are produced in the upper leaf axils in spring and early summer. The **fruit** is a woody follicle, 7–10cm long, containing 8–10 seeds.

Although this species has been in cultivation for decades, it has been wrongly labelled as *O. wickhamii* (now *Alloxylon wickhamii*). Most of the magnificent specimens seen in established parks and gardens are, in fact, this species, *A. flammeum*.

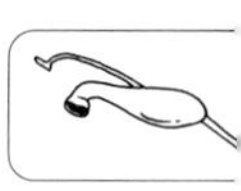

Silky Oak *Alloxylon wickhamii* to 30m

Although this beautiful flowering tree, formerly known as *Oreocallis wickhamii,* is fairly common in large areas of tropical rainforests around Cairns, north-eastern Queensland, it is rarely seen in cultivation. Until quite recently its scientific name was misapplied to the more widely cultivated *A. flammeum.* This Silky Oak occurs in mountainous areas between Mt Bartle Frere and Cape Tribulation, developing into a tall erect tree with a dense crown and a trunk diameter of up to 60cm. The **bark** is shallowly fissured and grey.

The leathery, ovate **leaves** are entire, 5–17cm long and up to 3.5cm wide. Immature leaves are pinkish-purple and have wavy margins. Its leathery leaves separate this species from *A. flammeum*, which has thin-textured, somewhat papery foliage. Dark pinkish-red **flowerheads** of 4–20 individual flowers are produced in terminal or axillary racemes in spring. These differ in colour from *A. flammeum*, which has bright orange-red flowers. The **fruit** is a woody capsule, 5.5–12cm long, containing 6–11 seeds.

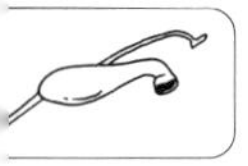

Coast Banksia *Banksia integrifolia* 5–25m

This common coastal tree shows a variation in height according to the exposure of its habitat, which may be on sand dunes very close to high tide level or a short way inland in open forests. It has a wide distribution in south-eastern Australia from around Fraser Island off Queensland, south to Tasmania.

Mature trees have a hard and roughly tesselated **bark**. Leathery **leaves** are in whorls and are narrow-obovate, 4–20cm long and up to 3.5cm wide. The margins are entire or occasionally with a few short teeth, and the undersides are silvery grey. Juvenile leaves are often coarsely toothed. The pale yellow **flowers** are produced in cylindrical spikes, 12cm long and 7cm wide, mainly in late summer to winter. Grey, cylindrical fruiting **cones** have prominent follicles that open wide at maturity, usually about 9 months after flowering. The old flowers are not persistent.

Because of its ability to withstand exposure to harsh sea winds, the coast banksia is an excellent shelter or windbreak tree in seaside gardens. It will also grow inland and is frost tolerant. The flowers are most attractive to honeyeaters and lorikeets. This is also a good honey-producing tree in autumn and early winter.

The ornamental pinkish-red timber has a distinctive pattern, polishes well and is easily worked. It is suitable for decorative work, such as panelling, cabinet work and turnery.

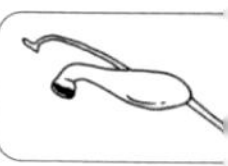

Saw Banksia *Banksia serrata* to 16m

This common banksia of the coastal areas of eastern Australia occurs from near Noosa in southern Queensland, south on the coastal plain and nearby ranges to south-eastern Victoria and north-western Tasmania. Saw Banksia sometimes occurs on exposed coastal heaths where it is a shrub, 1–3m tall, but mostly it is seen as a gnarled, often twisted, small tree with a dark grey warty **bark**, from which comes its alternate common name of old man banksia.

Its narrow-obovate **leaves**, 5–20cm long and up to 4cm wide, have evenly serrated margins and are usually green on both surfaces when mature. Creamy grey **flowers** are carried in large, broad, cylindrical spikes to 15cm long at the branch ends from summer to winter. The ribbed pollen presenter is spindle-shaped to 3mm long, and is the main botanical feature that separates this species from *B. aemula*, which has a smooth, short, conical pollen-presenter about 1mm long. Also, *B. aemula* usually has pale yellow (not greyish) flowers and its leaves arenarrower. The two species are occasionally found growing together.

Saw Banksia is noted for its wonderful massive **cones** of Snugglepot and Cuddlepie fame. The old flowers are persistent, giving a shaggy appearance, and the velvety follicles, up to 3.5cm long, are prominently exserted.

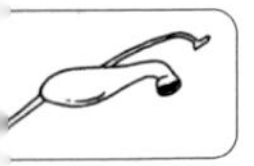

White Silky Oak *Grevillea hilliana* 8–30m

This prolific flowering tree grows within and at the edge of rainforests in coastal districts from the NSW/Queensland border north to the Cooktown area. It is a tall upright tree with angular branchlets and smooth grey **bark**. The **leaves** are very variable; the juvenile and intermediate leaves, 28–40cm long and 15–30cm wide, are deeply divided into 3–10 lobes. Most adult leaves are leathery, simple and oblong to obovate, 9–24cm long and 2–6cm wide, prominently veined and somewhat wavy. All of the leaves are dark green above with the lower surface silky grey.

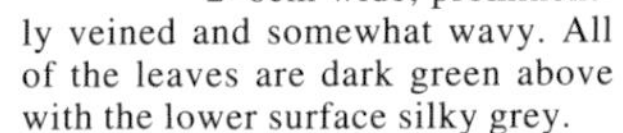

Creamy white **flowers** are produced in dense cylindrical racemes, up to 22cm long, mainly in winter. The racemes may be terminal or borne from the upper axils and are sometimes branched. The **fruit** is a leathery, ovoid follicle, about 2.5cm long.

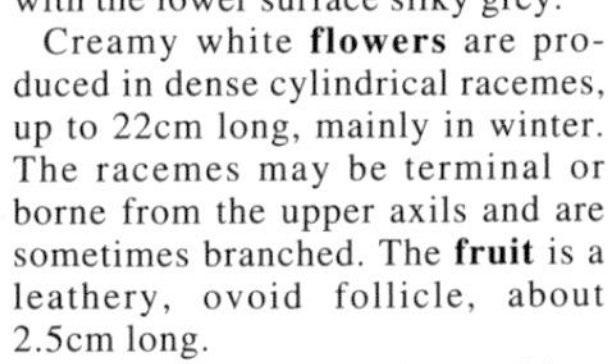

This beautiful garden subject forms a lower, more spreading tree when cultivated. It is long-lived and can be grown in temperate regions if given some protection from frost when young and regular watering during summer. The dark brown, close-grained timber is attractively patterned and has been used for cabinet work and veneers.

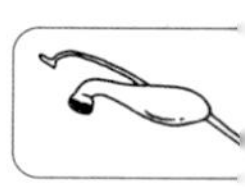

Silver Oak *Grevillea parallela* 3–14m

This slender weeping tree is widely distributed across tropical northern Australia in Western Australia, NT and Queensland. It can be seen in open forests, grasslands and on flat coastal plaints. The hard dark grey **bark** is furrowed. The species is characterised by its long, narrow, pendulous **leaves**, 10–40cm long, which are sometimes divided with linear lobes. The upper surface is dull green, the underside silky-hairy with 3–7 raised longitudinal veins. The white to cream, strongly fragrant **flowers** are produced in dense cylindrical racemes, 6–10cm long, arranged in groups of about 3, but sometimes as many as 7. Flowers appear in spring followed by orbicular to elliptic **fruit** to 3cm long and 2.5cm wide.

Golden Grevillea *Grevillea pteridifolia* to 10m

With its shimmering silvery appearance, this grevillea is attractive throughout the year. Also known as Fern-leaved Grevillea, it is widespread across tropical northern Australia in open forests and woodlands, often on floodplains or along streams. A prostrate form occurs on headlands around Cooktown, Queensland. Generally, it is a small, open-crowned slender tree with a silvery smooth **bark** that becomes furrowed with age. The pinnately lobed **leaves**, 25–45cm long, are grey-green above and silky and silvery below. Each long linear lobe is about 5mm wide. Bright orange or yellow-orange **flowers** are arranged on one-sided racemes up to 20cm long. The **fruit** is a somewhat flattened ovoid follicle, about 2cm long, covered in soft hairs. The nectar-rich flowers are used by the Aborigines to make a sweet drink.

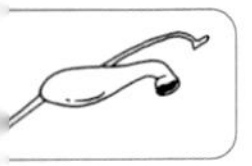

Silky Oak *Grevillea robusta* 8–40m

This large ornamental tree occurs naturally in coastal rainforests and inland ranges of far northern NSW, to about Maryborough, Queensland, and west to the Bunya Mountains. Tallest of the grevilleas, it has dark grey or nearly black, deeply fissured **bark**. It is easily recognised by its large deeply divided, fern-like **leaves**, 10–35cm long and 10–15cm wide, dark green above and whitish beneath. Bright orange **flowers** are produced in large one-sided racemes, 12–15cm long, in late spring and summer, followed by small boat-shaped **fruit**, to 2cm long, with 2 winged seeds. The strong, silken timber, durable and attractively marked, is used for cabinet-making and craftwork. Easily propagated from seed, it is grown throughout Australia and most countries of the world.

Beefwood *Grevillea striata* 8–15m

Beefwood is a single-stemmed, robust tree widely distributed in dry inland areas of all mainland states, except Victoria. It gets its common name from the deep red of the freshly cut timber. Attractively marked, this was used for cabinet work, furniture, fence posts and bush constructions. The tree has dark brown, rough, thickly furrowed **bark**. The flat linear, often curved, silvery grey **leaves**, 10–45cm long and mostly 1cm wide, have 5–15 raised parallel veins on the underside. White to cream **flowers** are produced in late spring and summer, usually in terminal, dense cylindrical racemes, 5–8cm long. The small, thin-walled, beaked **fruit** is less than 2cm long and wide.

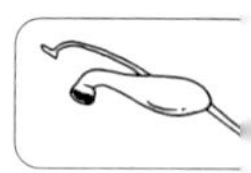

Bootlace Tree *Hakea chordophylla* 4–8m

This shapely and interesting inland tree is widespread in central Australia ranging from north-eastern WA to south-western Queensland. It has an open habit, contorted and pendulous branches and a corky furrowed **bark**. The extremely long, dark green, slender and cylindrical (terete) **leaves**, to 40cm, droop down from the branches. One could, indeed, use them for bootlaces.

The showy pendent **flowers,** in racemes up to 16cm long, are green at first opening to greenish-yellow. They are produced on the old wood, mainly throughout winter. The **fruit**, to 3.5cm long, tapers to a short beak.

Straggly Corkbark *Hakea eyreana* 4–7m

This small gnarled tree is restricted to a large inland area bordering four states: the extreme north-west of NSW, south-western Queensland, north-eastern SA and south-eastern NT. It has dark grey corky **bark**. The branchlets, leaves and flowers are woolly. The terete **leaves**, 3–10cm long, are divided several times into numerous pointed lobes. The yellowish **flowers**, borne mainly in winter and spring, appear in slightly pendulous racemes up to 10cm long. The **fruit**, to 4cm long, is ovoid and beaked. In traditional Aboriginal medicine, the bark is heated and ground to a fine powder and used as a treatment for burns.

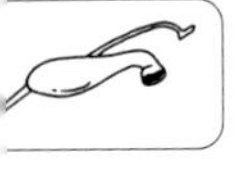

Corkbark *Hakea suberea* 2–9m

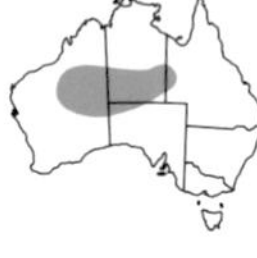

This small, picturesque gnarled tree is common in dry desert regions of the NT, SA and WA. It occurs on sandplains, limestone and gravelly soils, among rocks and on ridges. Its often twisted branches and trunk and thick, deeply furrowed dark brown corky **bark** give this tree its characteristic appearance. The needle-like, grey green **leaves**, 12–30cm long, are rarely divided. They are usually densely covered with whitish hairs and are often grooved beneath. The leaves usually point upwards or out, which separates this species from *H. chordophylla* with long, smooth, drooping leaves. This species differs from *H. divaricata,* another corkbark tree with a similar distribution, which has dull olive green leaves, 7–20cm long, divided into as many as 16 stiff segments.

Pendulous racemes of creamy yellow **flowers**, 4–15cm long, are produced terminally or in the upper axils in winter and spring. The slightly curved woody **fruit**, 3–4.5cm long and 2cm wide, has a prominent beak and sheds its winged seeds annually.

The honey-scented flowers, which produce copious quantities of nectar, are collected by Aboriginal women of central Australia and made into a sweet drink with water. Medicinally an infusion of the crushed bark is used as a wash for skin sores and scabies.

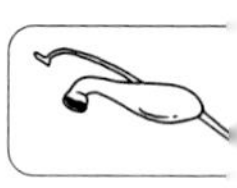

Red Boppel Nut *Hicksbeachia pinnatifolia* to 10m

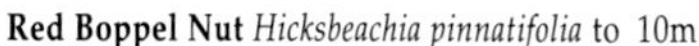

This erect, slender tree with bare lower branches grows in subtropical rainforests north of Lismore, NSW, to south-eastern Queensland. If it is cut or damaged, it often regenerates by suckering and may have multiple trunks in disturbed areas. The Red Bopple Nut has a smooth grey **bark** and features unusually large terminal pinnately lobed **leaves**, which are dark green and leathery with a paler undersurface. They are up to 1m long and 20cm wide and comprise 16–24 lobes, each 6–25cm long, with prickly toothed margins. Veins are prominent on both surfaces. Young growth, clothed in short, rusty brown hairs, is often an attractive purplish-red. Pinkish-purple **flowers** produced in pendulous racemes, 15–45cm long, appear only on the mature wood, often low on the main branches or arising from the trunk. They are heavily scented and are borne mostly in winter and early spring.

Shiny, bright red, oval **fruits**, 4cm long and 3cm wide, ripen in spring and summer and contain a single edible nut encased in a bony shell. With its large decorative leaves and colourful fruit, this is a striking small ornamental tree for the home garden in subtropical and warm temperate regions.

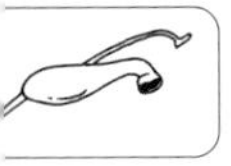

Macadamia Nut *Macadamia integrifolia* 8–15m

In its lowland rainforest habitat of south-eastern Queensland, this world-renowned fruit-bearing tree grows to medium size with a dense canopy. It has a much lower branching habit when cultivated in commercial nut plantations. The glossy, oblong to obovate **leaves**, mostly arranged in whorls of 3, are 7–15cm long and 2–4cm wide. They are entire with wavy or slightly prickly margins. White to pinkish **flowers** are borne in long pendent racemes, 10–20cm long, in winter and spring. The green leathery **fruits**, to 4cm across, have a hard, shiny inner shell encasing the sweet white kernel, which ripens in summer. The nuts are delicious raw or roasted.

Rough-shelled Bush Nut *Macadamia tetraphylla* 8–15m

This species is also commercially grown for its sweet edible nuts. It occurs naturally in coastal subtropical rainforests north of the Clarence River, NSW, to south-eastern Queensland. The dark green oblong to oblanceolate **leaves**, 6–30cm long, are mostly in whorls of 4 and have leaf margins with coarse spines. Pink or creamy pink **flowers** are produced in pendulous racemes, 10–25cm long, in late winter and spring. Greenish leathery **fruits**, about 3.5cm across, have a hard and rough inner shell encasing the sweet white nut, which ripens in summer and early autumn.

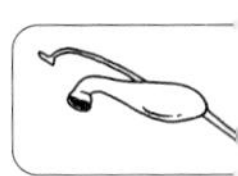

Firewheel Tree *Stenocarpus sinuatus* to 35m

A native of coastal rainforests north from the Nambucca River, NSW, to the Atherton District, Queensland, this well-known, handsome tree has been widely planted in parks and gardens for many years. It is noted for its umbels of spectacular orange-red **flowers** which, before they are fully open, resemble the spokes of a wheel. The flowers are produced on mature wood in autumn and winter and can be seen at branch ends or hidden inside the foliage. They are borne in great numbers and a tree in full flower is a magnificent sight.

The Firewheel Tree has a columnar habit with a dense crown of deep green foliage. The wrinkled or corky **bark** is dark grey or brown. Large leathery **leaves**, to 20cm long and 5cm wide, are variable in shape from deeply lobed to entire or shallowly lobed with wavy edges. They are shiny on the upper surface and pale beneath. The **fruit** is a brown, narrow follicle, about 10cm long, with a tapering tip and containing several overlapping, papery, winged seeds. Native birds love both the nectar-rich flowers and the seeds.

If young plants are protected from frost, this beautiful tree will thrive in most temperate climates. A sheltered position with deep, rich soil and ample moisture is preferred. Growth is slow and plants may take several years to flower.

Smooth-barked Apple *Angophora costata* 15–25m

This stately, straight-trunked tree, often mistaken for a eucalypt, can reach 25m or more in favourable conditions, such as in open forests, but in sandstone sites around Sydney it is short-trunked, gnarled and irregularly branched. It is one of the principal tree species in national parks near Sydney, where it may form almost pure stands and is locally known as the Sydney Red Gum. It is also widely distributed along the coast and ranges of NSW.

The deciduous **bark** sheds in large flakes in late spring, leaving a smooth, dimpled, pinkish-brown or bright orange surface that becomes grey with age. The bark is frequently stained a dark red by kino exudation (bottom photo). The lanceolate bright green **leaves**, 9–17cm long and 2–3.5cm wide, taper to a point. New leaves are bright red. Profuse creamy flowers are borne in clusters at the ends of the branches in late spring and summer. The individual flowers, about 2cm wide, have 5 tooth-like sepals, 5 petals and numerous stamens. The undeveloped flowers lack the operculum or bud cap that is a characteristic in the closely related eucalypts.

The slightly woody, goblet-shaped fruits, to 1.5cm long and across, are prominently ribbed. The tree is a valuable source of pollen and is prized by beekeepers. Both the foliage and bark have been used as dying materials.

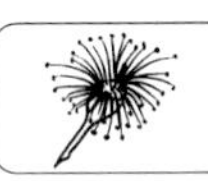

Rough-barked Apple *Angophora floribunda* to 30m

The widespread Rough-barked Apple, found in forested areas of the coast and ranges extending from far east Gippsland, Victoria, north to south-eastern Queensland, can also be seen scattered throughout the north-western slopes and plains of NSW. It is usually a medium-size to tall tree, frequently with low twisting branches and a spreading crown. The trunk is relatively short and covered with persistent grey-brown fibrous **bark**.

The lanceolate adult **leaves**, 8–12cm long and 2–3cm wide, taper to a point and are grey-green above and paler beneath. Profuse creamy white **flowers**, about 2cm across, are borne in spring and summer, usually in threes, in dense terminal clusters. Flowers produce copious nectar and are highly regarded as a source of honey. The thin-textured **fruits** are prominently ribbed and about 1cm in diameter.

The very similar *A. woodsiana*, a medium-size tree to 20m, found in northern NSW and southern Queensland, differs in having leaves to 14cm long and larger fruits, about 1.5cm in diameter. Angophoras are closely related botanically to the bloodwood group of eucalypts, so named because of the deep reddish-brown of the timber caused by the retention of kino within the gum veins in the wood. In 1995, bloodwoods were reclassified from *Eucalyptus* into a new genus, *Corymbia*.

Ghost Gum *Corymbia aparrerinja* to 20m

Ghost Gums around Alice Springs are among the most photographed, painted and publicised of inland Australian trees. They are widely distributed in central arid regions from the extreme east of WA through the NT to central Queensland. The species is famous for its beautiful, smooth, intensely white **bark**. Often contorted or gnarled from its struggle to survive under harsh conditions, its bright green, glossy **leaves** are narrow-lanceolate to lanceolate, often twisted, 5–16cm long and about 2cm wide. Pear-shaped **buds** have a hemispherical cap. White **flowers** are borne in umbels of up to 7, mainly in summer. Ovoid **fruits**, about 1cm long, are thin-textured and have enclosed valves.

In 1995, 'A revision of the bloodwoods of the genus *Corymbia* (Myrtaceae)' was published in *Telopea* Vol 6 (2–3). The work of K.D. Hilland and L.A.S. Johnson, it described 113 species of *Corymbia*, 33 for the first time, the remainder being reclassified from *Eucalyptus*. The new genus comprises those eucalypts traditionally known as bloodwoods and ghost gums.

The name *Eucalyptus papuana* has been widely and incorrectly applied to a number of ghost gums in northern and central Australia. It is now thought that the true *E. papuana* (now *Corymbia papuana*) is probably restricted to New Guinea and Australian ghost gums should be regarded as several distinct species. In the revision, *Corymbia aparrerinja* (formerly known as *E. papuana* var. *aparrerinja*) was described as a new species. *Aparrerinja* is the Aboriginal name used in central Australia for this species.

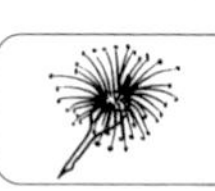

Smooth-stemmed Bloodwood *Corymbia bleeseri* 12–18m

This beautiful spreading tree is widespread in wetter parts of the NT and the Kimberley, WA, where it grows mostly in open forest in slightly hilly country. The **bark** is smooth and creamy white on the upper trunk and branches and scaly or flaky on the lower trunk, leaving blotchy rust-red and grey patches. The distinctive narrow-lanceolate **leaves**, 8–15cm long and 1–2.5cm wide, are highly glossy on the upper surface and only slightly paler beneath. The ovoid **buds**, about 1.2cm long, have a conical cap about the same length as the base. White **flowers**, to 2cm across, are borne in umbels of 7 on stalks to 3cm long, mainly in autumn. Urn-shaped **fruits**, to 2cm long, have deeply enclosed valves. Formerly *Eucalyptus bleeseri*.

Lemon-scented Gum *Corymbia citriodora* to 40m

Widely planted throughout Australia and overseas, this stately, medium to tall tree originates from Queensland, where it occurs mainly in dry open forests and woodlands south of Gladstone to the Maryborough district and up to 400km inland. It has a slender straight trunk with smooth, powdery white to grey **bark** throughout. Narrow-lanceolate adult **leaves**, 7–22cm long and about 2cm wide, have a sharp lemon fragrance when crushed. Small ovoid **buds** have a shortly pointed hemispherical cap. White **flowers** in umbels of 3 appear in summer and autumn. Urn-shaped **fruits,** to 1.5cm long, have deeply sunken valves. Formerly known as *Eucalyptus citriodora.*

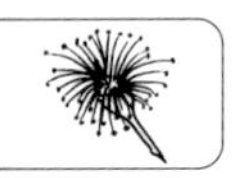

Red-flowering Gum *Corymbia ficifolia* 6–15m

One of the most beautiful and best-known of ornamental eucalypts, it is widely planted throughout the world but occurs naturally only on the coast west of Albany, WA. The tree is small or medium size with a shapely, compact crown and a short trunk with dark, greyish brown rough **bark** persistent on the upper branches. Mid-green, broad-lanceolate adult **leaves**, 7–14cm long and to 5cm wide, are smooth and leathery with a paler underside. The club-shaped **buds**, 2cm long, have a short hemispherical cap. Profuse **flowers** are borne in terminal clusters in summer. Colour varies from scarlet to crimson, pink, orange and white. The large, thick, urn-shaped **fruits**, 3.5cm long and 3cm wide, have enclosed valves. Formerly *Eucalyptus ficifolia*.

Red Bloodwood *Corymbia gummifera* to 30m

Under good conditions, Red Bloodwood is a medium-size to tall tree, but when things get tough, such as in coastal heaths, it may be reduced to a mallee. It is common in dry forests and woodlands mainly in a narrow coastal belt from south-eastern Queensland to north-eastern Victoria. The trunk is covered in short-fibred brown tessellated **bark** that persists on the small branches. The glossy dark green adult lanceolate **leaves**, 10–16cm long and 2–4cm wide, are paler on the undersurface. Ovoid flower buds, about 1cm long, have a short conical cap. Clusters of up to 7 creamy white **flowers**, each about 2cm across, are borne at branchends during summer. Woody, urn-shaped **fruits**, to 2cm long, have deeply enclosed valves. Formerly *Eucalyptus gummifera*.

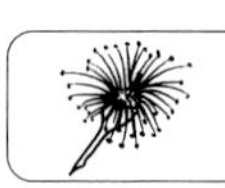

Leichhardt's Rusty Jacket *Corymbia leichhardtii* to 15m

This charming crooked tree occurs over a wide area on the slopes of the Great Dividing Range of central and northern Queensland. It is distinguished by the yellow-brown soft fibrous **bark**, which is persistent on trunk and branches. The ovate, peltate juvenile **leaves** and greyish-green, narrow-lanceolate to lanceolate adult leaves, 8–18cm long and about 2.5cm wide, separate this tree from similar species. The whitish ovoid **buds**, about 1cm long, have a rounded, shortly pointed cap, much shorter than the base. Honey-scented white **flowers** are carried in umbels of up to 7 in summer and autumn. The rounded **fruits**, about 1.5cm in diameter, have sunken valves. Formerly *Eucalyptus peltata* subsp. *leichhardtii*.

Spotted Gum *Corymbia maculata* 35–45m

This beautiful, tall, coastal tree extends from the Taree area south to near Bega, NSW, with an isolated occurrence south of Buchan, Victoria. It has a long straight trunk and smooth pale grey, pink or cream deciduous **bark** that is shed in irregular patches, leaving slight dimples on the surface. Narrow-lanceolate adult **leaves**, to 22cm long and 4cm wide, are dull green on both surfaces. Ovoid **buds**, to 1cm long, have a short hemispherical cap with a blunt tip. White **flowers** in umbels of 3 appear mostly in autumn and winter. The ovoid to urn-shaped **fruits**, to 1.5cm long, have deeply enclosed valves. Spotted Gum, formerly *Eucalyptus maculata*, closely resembles *C. citriodora*, but the essential oil in the leaves lacks the distinctive lemony odour of the latter.

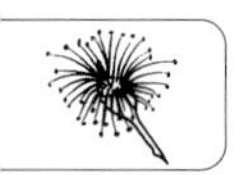

Inland Bloodwood *Corymbia opaca* 8–15m

Widely distributed in the NT, this inland plains tree, formerly *Eucalyptus opaca*, occurs in deserts of central WA, east to the Simpson Desert and north to the Kimberley region. Small to medium size, with a slender crooked trunk, it sometimes has several stems. The pale brown to orange-brown tessellated **bark**, persistent on the trunk and large branches, may be smooth and creamy-coloured on the upper parts of the tree. Lanceolate adult **leaves**, 9–17cm long and to 3.5cm wide, are dull green on both surfaces.

The pale brown ovoid **buds** have a shortly pointed rounded cap much shorter than the base. Profuse white **flowers** are carried in umbels of up to 7 during winter. Ovoid **fruits**, to 3cm long, are widest below the middle and have deeply enclosed valves.

Carbeen *Corymbia tessellaris* to 30m

Widely distributed over most of eastern and central Queensland and into northern NSW, this distinctive tall tree also occurs in south-western Papua New Guinea. It has a stocking of dark grey tessellated **bark** persistent over about half the trunk, with the upper trunk and branches smooth and white or pale grey. The narrow-lanceolate to lanceolate adult **leaves**, 12–18cm long and about 2cm wide, are green to grey-green on both surfaces. The small pear-shaped **buds** have a dome-shaped cap. White **flowers** in umbels of up to 7 appear in spring or summer, depending on its range. Thin-walled, ovoid **fruits** to 1.2cm long have deeply enclosed valves. Formerly *Eucalyptus tessellaris*.

White Gum *Eucalyptus alba* 15–20m

A striking medium-size tree of the Top End of the NT, noted for its beautiful smooth white **bark**. It occurs mostly along freshwater streams, floodplains, in seasonally wet depressions and other moist sites. The dull green lanceolate **leaves**, 10–20cm long and 1.5–6cm wide, taper to an elongated tip. Small ovoid **buds**, about 6mm long and 4mm across, have a rounded to slightly pointed cap. Small creamy white **flowers**, in clusters of up to 7, are borne in the leaf axils in late winter and early spring. The woody cup-shaped **fruits**, about 6mm across, have 3 or 4 slightly protruding valves. Formerly known as *E. alba* var. *australasica*.

Northern Salmon Gum *Eucalyptus bigalerita* 7–14m

The main habitat of this outstanding tree is along watercourses or low-lying flats subject to flooding during the wet season in the north-western NT and the north of the Kimberley region, WA. The smooth white **bark** is shed annually exposing cream or salmon bark beneath. The long-stalked, thick ovate to broadly ovate **leaves**, 6–15cm long and 5–13cm wide, are usually glossy green on both sides. Almost rounded **buds**, about 1cm long and across, have dome-shaped caps. Profuse white flowers, about 1.5cm wide, are produced in axillary umbels of 7 in winter and spring. Cup-shaped **fruits**, about 1cm long and wide, have 4 slightly protruding valves. A similar species, *E. tintinnans*, is found south-east of Darwin on gravelly hillsides and stony ridges. It has salmon-coloured bark, smaller dull green leaves and smaller buds and fruit.

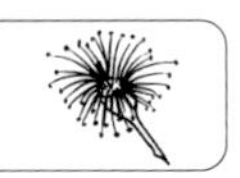

River Red Gum *Eucalyptus camaldulensis* 15–30m

This charming, stately tree with a spreading, open crown grows along rivers, creeks and dry watercourses throughout inland Australia. It is the most widely distributed of all the eucalypts and is widely cultivated throughout the world for its ornamental value, rehabilitation and soil stabilisation work and for fuelwood and charcoal production.

The single or multiple, often massive, trunks are sometimes twisted or contorted by nature and have smooth, attractively mottled **bark** with white, grey, blue, cream and reddish patches. Pendent adult **leaves** are lanceolate to narrow-lanceolate and are variable in length, 10–25cm long and to 2cm wide. Small **buds**, 1cm long and 50mm wide, have a strongly beaked cap longer than the base. Profuse white **flowers**, in axillary umbels of 7–11, occur mainly in late spring and summer. Small hemispherical **fruits**, about 1cm across, have 3–5 prominent exserted triangular valves. When mature trees shed their lower branches, hollows are created that become important nesting sites for large birds and dens for arboreal mammals and reptiles.

In Aboriginal medicine, young soft leaves are mixed with equal amounts of water and applied as a wash for fever, headache and to treat coughs and colds. The hard, reddish timber is resistant to termites and decay and is used for heavy construction, poles, railway sleepers, fencing and flooring. Flowers are rich in nectar and the tree is highly regarded for honey production.

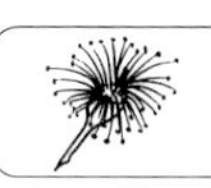

Karri *Eucalyptus diversicolor* to 70m or more

The magnificent Karri is the tallest tree found in WA and one of the 3 tallest hardwood species in the world. It occurs in higher-rainfall regions of the extreme south-west, extending roughly from around Margaret River to Albany, where it is the dominant tree of the important Karri forests. Fine examples of old-growth Karri forest can be seen at Warren National Park, near Pemberton.

This forest giant has a straight trunk with smooth **bark**, in tones of whitish-grey or yellowish-brown, which sheds in irregular blotches. Juvenile **leaves** are ovate to almost rounded and about 10cm in diameter. The adult leaves are broad-lanceolate, 9–12cm long and about 3cm wide. They are dark green above and distinctly paler on the underside. The botanical name *diversicolor*, which means separate colours, alludes to the colour difference between the upper and lower surface of the leaves. Club-shaped **buds**, about 1.5cm long, have a short conical cap. Creamy white **flowers** are borne in axillary clusters of up to 7 in spring and summer. Ovoid **fruits**, about 1cm in diameter, have enclosed valves.

The very hard reddish-brown timber is highly regarded and is used for building purposes and by the woodchip industry. Karri is also an important honey-producing tree.

Flooded Gum *Eucalyptus grandis* 45–55m

In coastal districts north from Newcastle, NSW, to south-eastern Queensland, Flooded Gum is common in wet sclerophyll forests and bordering rainforests. Smaller populations also occur to the west of Mackay and again on the Atherton Tableland, Queensland.

A tall tree with a straight shaft-like trunk, it has a short stocking of persistent fibrous **bark** at the base and smooth powdery white bark above. Glossy dark green adult **leaves** are lanceolate, 10–16cm long and 3cm wide, and have a paler underside. The small ovoid **buds** have a conical cap. White **flowers**, about 2cm across, are borne in axillary umbels of 7–11 in autumn and winter. The pear-shaped **fruits** have 4–5 broad, slightly exserted, incurved valves and are often glaucous, which distinguishes this species from the closely related Sydney Blue Gum, *E. saligna,* which has smooth fruits with thin pointed valves that curve outwards.

Flooded Gum is a major timber-producing tree. The strong, pinkish-brown wood is used for house construction, panelling, flooring, plywood and veneers. One of the fasting growing eucalypts, it is a successful plantation forestry tree in Australia, South Africa and California.

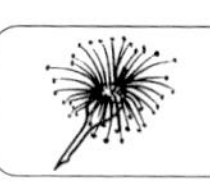

Broad-leaved Scribbly Gum *Eucalyptus haemastoma* to 15m

This lovely tree flourishes on poor shallow soil on sandstone and is a feature of the bushland around Sydney. In some areas it is a medium-size tree with a solitary trunk and fairly dense canopy, but on infertile soils especially, it sometimes takes a mallee-like habit. The frequently short crooked trunk has white, yellowish or pale grey satiny **bark** marked by prominent irregular scribbles. These are caused by insect larvae burrowing beneath the old bark before it is shed.

Juvenile **leaves** are ovate, dull grey-green, and about 8cm wide. The thick, broad-lanceolate adult **leaves**, 12–15cm long and to 4cm wide, are curved and glossy green on both surfaces. These are an important food source for koalas. **Buds**, to 1.5cm long and 1cm across, are club-shaped with a short rounded cap. Small white **flowers** are formed in axillary or sometimes terminal umbels of up to 12 in autumn and spring. Conical or pear-shaped **fruits**, about 1cm in diameter, have a thick, reddish rim on top. The valves are slightly enclosed or at rim level.

Other species of NSW Scribbly Gum are the Hard-leaved Scribbly Gum, *E. sclerophylla*, found north and south of Sydney; the Narrow-leaved Scribbly Gum, *E. racemosa*, confined to the central coast south of the Hunter River, and the Inland Scribbly Gum, *E. rossii*, on the tablelands. *E. haemastoma* is distinguished from these by its much broader leaves, larger buds and fruit.

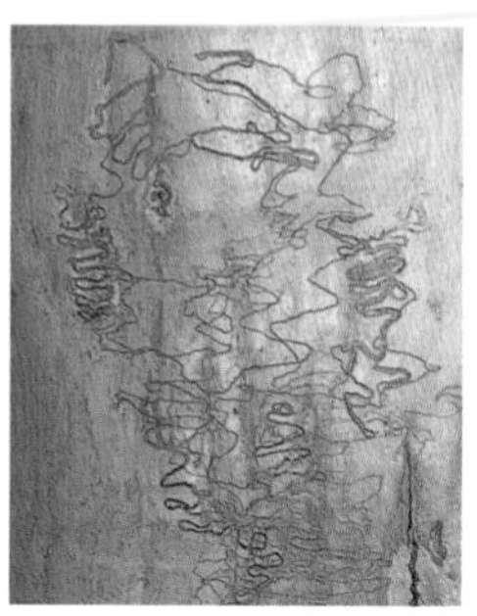

Yellow Gum *Eucalyptus leucoxylon* 10–30m

Usually a small or medium-size woodland tree with a single trunk, but on drier sites Yellow Gum can be reduced to a mallee. It is fairly common in central and western Victoria in dry forest communities dominated by box and ironbark eucalypts, but it also occurs in south-eastern SA and south-western NSW. Sometimes there is a short stocking of fibrous **bark** at the base; the remainder is shed in irregular flakes, leaving a smooth, creamy yellow or bluish-grey surface.

Adult **leaves**, 7–20cm long and to 3.5cm wide, are narrow lanceolate, slightly curved, grey-green on both sides and with a prominent central vein. They hang vertically in a fairly open crown. Ovoid **buds**, to 1.5cm long, have a short rounded cap about the same length as the base. Profuse **flowers**, about 3cm across, are borne in pendulous clusters of 3 from late autumn to spring. They are usually white or cream, but also pink, red or yellow. The **fruits** are mostly cup- or barrel-shaped, to 1.2cm in diameter with enclosed valves. Subspecies *megalocarpa* has deep reddish-pink flowers and is the form most widely planted as an ornamental. Its large, bell-shaped ribbed fruits are up to 1.8cm long and 1.6cm wide. Some forms of subspecies *petiolaris*, restricted to south-eastern Eyre Peninsula, SA, have bright yellow flowers.

Easily grown in most temperate and semi-arid areas, the adaptable Yellow Gum will withstand wind, drought and frost. It is highly regarded for honey production and the pale, yellowish-brown, strong, durable wood is used for buildings, railway sleepers, poles and fences.

Snow Gum *Eucalyptus pauciflora* to 20m

A beautiful cold-country species, usually less than 20m, it is widely distributed and is locally common in mountains and tablelands of NSW, Victoria and Tasmania. It has a short trunk and deciduous **bark** shed in irregular patches leaving a smooth, white, grey or yellowish surface. The shiny lanceolate adult **leaves**, 14–18cm long and up to 4cm wide, have distinctive veins running almost parallel to the midrib. Club-shaped **buds**, to 1cm long, have a rounded conical cap. Numerous white **flowers** are borne in axillary clusters of 15 or more in late spring or summer. Cup-shaped **fruits**, about 1cm across, have thick rims and slightly enclosed valves. Subspecies *niphophila* (in photograph), with a mallee-like habit, is confined to high peaks in alpine areas. Its buds, fruit and leaves are glaucous.

Candlebark *Eucalyptus rubida* 30–40m

This medium to tall tree of open forests and woodlands ranges from the northern tablelands of NSW to south-eastern Tasmania, with isolated populations in the Mount Lofty Range, SA. It has smooth, white, greyish-pink or red **bark** that is shed from the whole tree in large strips. Narrow-lanceolate adult **leaves**, 9–15cm long and to 2.5cm wide, are dull green or grey-green. Small, ovoid **buds** have a conical cap shorter than the base. White **flowers** are borne in axillary umbels of 3 in summer and autumn. Hemispherical **fruits**, about 7mm in diameter, have a slightly raised broad disc and exserted valves. Young growth, juvenile leaves, buds and occasionally fruits are covered by a waxy white bloom.

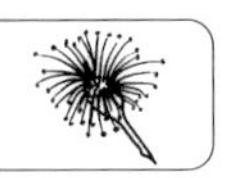

Sydney Blue Gum *Eucalyptus saligna* 30–50m

Fairly common in coastal areas from Maryborough, Queensland, to Batemans Bay, NSW, this beautiful forest tree grows mostly in wet sclerophyll forests. It has a straight trunk with a low stocking of rough **bark** and a smooth, satiny bluish-grey to white bark above. The bark and other features resemble that of the closely related *E. grandis*, which differs only in the glaucous buds and fruits and the broad incurved valves of the fruit. The glossy adult **leaves** are broad-lanceolate, 10–17cm long to 3cm wide, with paler underside and prominent veins. Ovoid **buds**, to 8mm long, have a slightly beaked cap, as long and wide as the base. White **flowers**, 2cm across, are borne in axillary umbels of 7–11 in summer and autumn. The pear-shaped **fruits**, to 8mm long, have 3 or 4 narrow, pointed, outward-curving valves protruding above the rim.

Salmon Gum *Eucalyptus salmonophloia* 15–30m

A beautiful, medium to tall tree with a shiny open crown and a smooth, satiny, salmon-red **bark** when fresh, weathering to brown-grey and shedding in flakes. It is widespread in the low-rainfall areas in the wheatbelt and goldfields of WA and is often a conspicuous tree in open forests and woodlands. Bright glossy green adult **leaves** are lanceolate, 7–12cm long and 1.5cm wide. Yellowish-green club-shaped **buds** have a rounded cap slightly longer than the base. Small white to cream **flowers** are produced in axillary clusters of up to 13, mainly in summer. The hemispherical **fruits**, about 50mm in diameter, have slender exserted valves. The strong durable timber was once used extensively in the mining industry.

Forest Red Gum *Eucalyptus tereticornis* 20–50m

Forest Red Gum extends north from Gippsland, Victoria, along the coastal side of the Divide to north of Cairns, Queensland, and into southern Papua New Guinea. This beautiful large tree is mostly seen in open forests and on alluvial flats or granite hillslopes. The **bark** is usually shed from the whole trunk and branches in large irregular sheets, leaving a smooth, mottled surface in delicate shades of cream, bluish-grey and white. Adult **leaves** are lanceolate, 10–20cm long and 2.5cm wide, and shiny green on both sides. **Buds**, to 2cm long, have horn-shaped caps longer than the base. Profuse white **flowers** in axillary umbels of 7–11 occur in winter and spring. The ovoid **fruits** have a raised disc and 4 or 5 strongly exserted valves.

Coral Gum *Eucalyptus torquata* 6–12m

Widely known in cultivation, this attractive, small spreading tree can be seen growing naturally on stony hills on red soil along the roadside from Norseman to Kalgoorlie, WA. It has rough grey **bark** on the trunk and smooth grey branches above. Young stems are red. Adult **leaves** are grey-green, lanceolate and slightly curved, to 12cm long. The red, ribbed **buds**, to 2.5cm long, have a distinctive pointed cap slightly longer than the base. Pendulous **flowers** in shades of pink or red, sometimes white or cream, are borne in axillary clusters of up to 7 in spring and summer. Cylindrical **fruits**, to 1.5cm long, have a ribbed base and mostly enclosed valves.

Manna Gum *Eucalyptus viminalis* to 45m

Also known as Ribbon Gum, this tall forest tree has a stocking of rough **bark**; the remainder is shed in long ribbons, leaving a smooth white or yellowish upper trunk and branches. Often the bark remains hanging untidily in branch forks. Manna Gum is widely distributed in southern Tasmania, from SA to Victoria and along the NSW tablelands to the Queensland border. Dark green adult **leaves** are lanceolate, often curved, 10–20cm long and 2cm wide. They are eaten by koalas. Ovoid **buds**, 1cm long and 55mm wide, have a conical cap as long and as wide as the base. White **flowers**, about 2cm across, are produced in axillary umbels, mostly of 3, in summer and autumn. Globular **fruits**, about 1cm across, have a raised disc and 3 or 4 exserted valves.

Wandoo *Eucalyptus wandoo* 8–30m

Wandoo, a medium-size tree of south-west WA, is found mainly on the western edge of the wheatbelt from Mingenew to Albany. It has smooth, mottled, yellow and white **bark** and powdery grey branches. The dull green adult **leaves** are narrow-lanceolate, 8–12cm long and to 2cm wide. Horn-shaped **buds** have an elongated conical cap about twice as long as the base. Profuse cream **flowers**, about 3.5cm across, are borne in showy axillary racemes of up to 15 from late spring through to autumn. Cylindrical to pear-shaped **fruits**, about 1cm long, have level or sometimes slightly exserted valves. Wandoo is highly regarded as a honey producer and noted for its strong, durable timber used in all forms of heavy construction work.

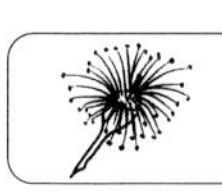

Brown Stringybark *Eucalyptus baxteri* to 40m

Common in south-western Victoria and south-eastern SA, Brown Stringybark extends to eastern Victorian forests and the south-eastern corner of NSW. It is very variable in size and can be a tall tree in forests or mallee-like in exposed coastal conditions. The brown **bark** is thick, fibrous and stringy and persistent to the small branches. The thick and glossy green adult **leaves** are broad-lanceolate, 7–12cm long and to 3cm wide. Club-shaped **buds** have rough blunt caps shorter than their bases. Profuse white **flowers**, to 2cm across, are borne in axillary clusters of 7–15, mostly in summer. The broad rounded **fruits**, about 1cm long and 1.6cm wide, have a broad disc and 4 or 5 slightly exserted valves.

Narrow-leaved Red Ironbark *Eucalyptus crebra* 20–30m

This medium-size woodland tree has a wide coastal distribution to just west of the Great Dividing Range from far northern Queensland to south of Sydney. Coarsely furrowed, blackish **bark**, persistent to the small branches, is often impregnated with kino. Narrow-lanceolate adult **leaves**, 7–15cm long and to 1.5cm wide, are dull grey-green on both sides. Club-shaped **buds**, about 6mm long and 4mm wide, have a conical cap about as long as the base. Clusters of white **flowers** are borne in umbels of 7–11 at branch ends from May to January. Small **fruits** are cup- or barrel-shaped, about 7mm long and 6mm wide, with valves at or below rim level. An excellent commercial tree, its very hard, strong, durable timber is suitable for heavy construction and railway sleepers.

Dundas Blackbutt *Eucalyptus dundasi* 9–20m

This distinctive medium-size tree has a limited distribution but is common around Norseman, WA, and is easily recognised by the rough, dark, tessellated **bark** on the lower part of the trunk. Above this, the smooth bark in late summer is richly copper-coloured, aging to grey. The adult **leaves**, 8–10cm long and 1.5cm wide, are lanceolate and glossy green on both sides. Small ovoid **buds** have a hemispherical cap with a prominent horn-shaped tip. Showy creamy white **flowers**, 2cm across, are borne in umbels of up to 7 in autumn. Cylindrical **fruits**, 1cm long, have 2 fine ribs and deeply sunken valves. This tree was named after Dundas, an early mining town where it was found.

River Peppermint *Eucalyptus elata* 20–30m

A slender tree found on eastern parts of the coastal ranges south from Putty, NSW, to East Gippsland, Victoria. It occurs in open forests, most commonly along the banks of rivers and in small valleys. Grey to blackish **bark** persists on the lower trunk, with the upper trunk and branches shedding annually in long hanging ribbons. Thin-textured narrow-lanceolate adult **leaves**, 9–15cm long and about 1.5cm wide, taper to a long point, and have numerous oil glands with a eucalyptus-menthol smell when crushed. The small club-shaped buds have a short hemispherical cap, shorter and narrower than the base. Up to 40 creamy white flowers, to 1.5cm across, are borne in large axillary clusters in spring. The small globular fruits, about 5mm across, with slightly sunken valves, are also in large clusters.

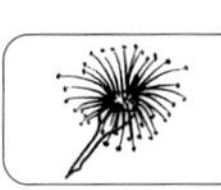

Broad-leaved Red Ironbark *Eucalyptus fibrosa* 15–35m

This medium to large tree is widespread along the coast and slightly inland from southern NSW north to Rockhampton, Queensland. It frequently has a large crown and dark brown to black, deeply furrowed **bark** persistent to the smaller branches. The adult **leaves**, 12–19cm long and to 5cm wide, are broadly lanceolate, leathery and grey-green on both surfaces. Elongated **buds**, to 2cm long, have a slender pointed cap longer than the base. Profuse white **flowers**, to 2cm across, are borne in umbels of 7–11 in late spring and summer. The pear-shaped **fruits**, to about 1cm in diameter, have 4 or 5 valves at rim level or exserted. The dark red wood, hard and extremely durable, is used for heavy construction and railway sleepers.

Black Box *Eucalyptus largiflorens* 10–20m

Occurring over most of inland NSW, extending into Queensland, Victoria and SA, this shapely tree is often dominant in open woodlands in low-lying ground associated with periodic flooding. It has a short, often crooked trunk and grey to black fibrous **bark** persistent to the smaller branches. The dull green, narrow-lanceolate **leaves**, 6–15cm long and less than 2cm wide, often have a curved tip. Ovoid **buds**, 5mm long, have a small conical cap shorter than the base. Creamy white **flowers** are borne in axillary or terminal umbels of up to about 7 in spring and summer. Hemispherical fruit, about 5mm in diameter, has 3–4 tiny valves just below rim level. The nectar-rich flowers produce good-quality honey.

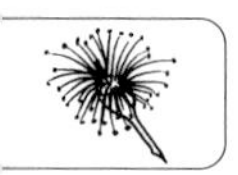

Red Stringybark *Eucalyptus macrorhyncha* to 30m, but mostly less than 20m

This straight-trunked species is widespread in woodlands, mainly on the foothills and lower slopes of the Great Dividing Range from northern NSW to western Victoria, with an isolated occurrence in SA. Its reddish-brown stringy **bark** is persistent throughout and weathers to dark brown or grey. The lanceolate adult **leaves** are glossy green on both sides, 9–14cm long and up to 3cm wide. Diamond-shaped buds have a pointed beaked cap about as long as the base. Umbels of 7 or more white **flowers**, about 1. 5cm across, are carried on a flattened stalk in summer and early autumn. The **fruits** are rounded, about 1.2cm across, with a broad, domed disc and 3 protruding valves.

Jarrah *Eucalyptus marginata* 30–40m

One of Australia's most important hardwoods, Jarrah grows in pure or almost pure stands in woodland areas in the south-west corner of WA. The strong, dark red or reddish-brown timber is used extensively for heavy construction, in the building industry and for furniture making. The trunk is straight with rough, reddish-brown to grey fibrous **bark**. Dark green broad-lanceolate to lanceolate adult **leaves**, 8–13cm long and 3cm wide, taper to a fine point. The elongated **bud** has a narrow, horn-shaped cap, often reddish, that is much longer than the base. An important honey-producing tree, its white to cream **flowers** in axillary clusters of 7 to 11 appear in spring. The globular **fruits** are about 1.5cm in diameter.

Yellow Box *Eucalyptus melliodora* 15–30m

The botanical name of this famous honey-producing tree is derived from the Latin *melleus* meaning honey and *odora*, sweet or pleasant smell, referring to the fragrance of the honey-laden blossoms. An alternate common name is Honey Box. Yellow Box is a medium to tall tree widespread and common on tablelands and western slopes from south-eastern Queensland to western Victoria. It usually grows in woodlands on moderately fertile soil in association with various other eucalypts.

The **bark** is variable, but usually rough and fibrous or flaky on the trunk and branches and smooth white, cream or grey on the upper parts. Occasionally, the bark is shed throughout, leaving smooth irregular streaks of dark and light grey. Adult **leaves**, 6–14cm long and to 1.8cm wide, are narrow-lanceolate or lanceolate and greyish-green on both sides. The diamond-shaped **buds**, 8mm long and 4mm wide, have a conical cap shorter than the base with no seam at the join. Profuse, sweetly scented white or, rarely, pink **flowers** are borne in axillary umbels of 7 in spring and throughout summer. Ovoid or hemispherical **fruits**, about 7mm in diameter, have enclosed valves. Yellow Box is prized as an ornamental and shade tree in rural areas and is regarded by many as one of the best honey producers in Australia. The mostly persistent bark and umbels of 7 flowers separate it from the closely related smooth-barked *E. leucoxylon*, which has larger buds and fruit.

Western Grey Box *Eucalyptus microcarpa* to 25m

Occurring mainly in woodlands, this species is widespread and common on slopes and plains inland of the Great Dividing Range from central Queensland through to south-western Victoria, with isolated occurrences in SA. It has a straight trunk usually about half of the tree's height, but in poorer soils it might be a smaller tree up to 10m tall. The rough, finely fissured grey **bark** is persistent on the trunk and larger lower branches, becoming smooth on the smaller upper branches.

Narrow-lanceolate adult **leaves**, 7–15cm long and to 2.5cm wide, taper at both ends and are usually dull green on both surfaces. Juvenile leaves are ovate. The small ovoid **buds** have a conical cap, often with a curved tip. The white **flowers**, to 1.5cm across, are borne in clusters of 7–9, mostly in autumn. Small barrel-shaped **fruits**, about 7mm long and 5mm across, have 4 valves slightly enclosed. The strong and very durable light brown timber is used for construction, fence posts and railway sleepers and is highly valued as firewood. The tree is also an excellent honey producer.

The closely related Grey Box, *E. moluccana*, has broader adult leaves, larger buds and fruits and persistent bark only on the trunk. The White Box, *E. albens*, differs in having glaucous buds and fruits, greyish-green adult leaves and a whitish-grey box-type bark.

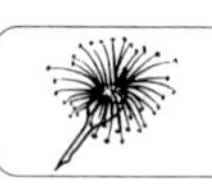

Tallowwood *Eucalyptus microcorys* 35–60m

This distinctive tree is widely distributed in wet sclerophyll forests on rainforest margins in coastal areas from around Newcastle, NSW, to Maryborough, Queensland. It is tall, usually around 40m, but occasionally reaching 60m. It is recognised by its persistent, soft, reddish-brown fibrous **bark**, often with surface pores, on the trunk and small branches. It has a horizontal branching habit and glossy dark green lanceolate adult **leaves** 8–12cm long and to 2.5cm wide. Club-shaped **buds**, about 6mm long and 3mm wide, have a small rounded cap with a faint cross-like marking. Creamy white **flowers**, to 1.5cm across, are borne in showy terminal clusters of up to 11 from winter to early summer. Cone-shaped or pear-shaped **fruits**, to 1cm long and 6mm across, have 3 or 4 valves at rim-level.

Tallowwood produces one of the best hardwoods in Australia. It is extremely strong and durable and is used for heavy construction work as well as external use such as window sills. The common name refers to the greasy nature of the yellow-brown timber, which, when dressed, provides a particularly popular flooring timber.

This is an excellent shade and shelter tree for homesteads, parks and public gardens in frost-free districts. Koalas feed on the foliage.

Coolibah *Eucalyptus microtheca* 10–20m

A widely distributed eucalypt found across northern Australia, the Coolibah grows in open grassy woodlands on seasonally flooded poorly drained areas or near larger rivers. It is small to medium size, with a fairly short or crooked trunk and spreading gnarled branches. It is sometimes taller and straighter in areas receiving regular floods.

The fibrous, flaky grey **bark** is persistent on the trunk and main branches; the smaller upper branches are smooth and whitish-grey. The narrow-lanceolate to lanceolate **leaves**, 8–17cm long and 1–2.5cm wide, are dull green or grey green. The tiny ovoid **buds,** about 3mm long, have a short, rounded cap. White **flowers**, about 1cm across, are borne in umbels of up to 11 that form part of larger clusters. Flowers appear mostly in summer and autumn. The small hemispherical **fruits**, about 4mm in diameter, have mostly enclosed valves. *Microtheca* means 'small fruit'.

In Aboriginal medicine, a wash is made from the leaves or inner bark and applied to the body for sores, aching joints and symptoms of cold and flu. A poultice is also made from pounded inner bark and used to treat snakebite. The seeds, which are first soaked, dried and ground, are made into an edible paste and the roots can be tapped for water.

Darwin Woollybutt *Eucalyptus miniata* 10–20m

A widespread and dominant tree of tropical woodlands of northern Australia extending from the Kimberley region, WA, across the Top End of the NT and the southern part of Cape York Peninsula, Queensland, it often grows in association with a number of other eucalypts, including Darwin Stringybark, *E. tetrodonta*.

Darwin Woollybutt is an erect tree with an open spreading crown. The dark grey to almost black fibrous **bark** is persistent for about half the trunk, then smooth, creamy white and powdery on the upper trunk and branches. The smooth, lanceolate to broadly lanceolate adult **leaves**, 7.5–15cm long and 5cm wide, have distinct lateral veins, an elongated tip and are light green above and paler below. The club-shaped **buds,** with a short cap, are ribbed and covered in whitish bloom. Spectacular orange-red **flowers**, to 3cm across, are borne in umbels of 3–7 on short thick stalks from May to August. The urn-shaped **fruits**, to 6cm long and 5cm across, have prominent ribs and deeply sunken valves.

The nectar-rich flowers attract fruit bats and many birds. It is an important honey tree for Aborigines. Medicinally, an infusion of the inner bark is taken for the treatment of diarrhoea and used as a lotion for small wounds and swellings.

New England Peppermint *Eucalyptus nova-anglica* 12–25m

This tree is common on the New England tablelands of northern NSW and extends into southern Queensland. Its short, fibrous grey-brown **bark** is persistent on the trunk and larger branches and smooth and shedding on smaller branches. Adult **leaves** are narrow-lanceolate to lanceolate, 7–15cm long and up to 1.3cm wide, and bluish-grey on both surfaces. The rounded, silvery blue juvenile leaves are often retained. The frosted ovoid **buds** have a blunt conical cap. White **flowers**, about 1cm across, are produced in umbels of 4–7 in summer and autumn. The small conical **fruits** have a raised disc and exserted valves.

Messmate Stringybark *Eucalyptus obliqua* 45–90m

This large, sometimes very tall tree with rough fibrous **bark** is widely distributed in cooler southern parts of eastern Australia from southern Queensland to south-eastern SA as well as Tasmania. In near coastal situations or in drier areas, it may be stunted or mallee-like. The curved, asymmetrical, dark green adult **leaves** are broad-lanceolate, 10–15cm long to 3cm wide, and have distinct veins. The club-shaped **buds** have a rounded pointed cap much shorter than the base. Profuse creamy white **flowers** are borne in umbels of 15 or more in summer and autumn. Barrel-shaped **fruits**, about 1cm in diameter, have a descending disc and enclosed valves. The timber is one of the most important hardwoods in Australia, used for construction purposes and pulp production.

Blue Mountains Ash *Eucalyptus oreades* 30–40m

This tall straight tree of higher altitudes is common in NSW in the Blue Mountains and inland from Port Macquarie as well as on both sides of the NSW/Queensland border. It has smooth white or yellowish **bark** that is shed in long, narrow ribbons, leaving a short stocking of rough bark at the base. Curved adult **leaves** are lanceolate, 11–17cm long and to 2.5cm wide, and dark green on both surfaces.

Club-shaped **buds** have a pointed conical cap about as long as the base. The white **flowers** in axillary clusters of 7 are produced in summer. The short-stalked, barrel-shaped **fruits**, about 1cm in diameter, have a descending disc and slightly enclosed valves.

Blackbutt *Eucalyptus pilularis* to 70m

A medium to tall tree, common in coastal wet and dry sclerophyll forests from near Gympie in south-eastern Queensland to southern NSW, it has rough, persistent grey **bark** on most of the trunk. Deciduous bark is shed in long strips on the upper trunk and branches leaving a smooth cream or greenish surface. The curved lanceolate adult **leaves** are 10–16cm long and 4cm wide, dark green above and paler below. Spindle-shaped **buds** have a pointed conical cap longer than the base. White **flowers** are produced in axillary clusters of 7–15, mainly in spring and summer. Hemispherical **fruits** have valves at rim level or slightly enclosed.

Other tall eucalypts with a rough stocking of bark blackened by fire are commonly called Blackbutt.

Mountain Ash *Eucalyptus regnans* 55–80m

The tallest species of *Eucalyptus* and the tallest hardwood in the world, Mountain Ash often grows in impressive stands in tall open forests in mountainous areas of eastern Victoria and Tasmania where high rainfall produces trees of immense height and girth. Measurements of up to 100m have been recorded.

It is an erect, straight tree with rough persistent **bark** 5–20m from the base; the remainder is shed in long ribbons, revealing a smooth whitish or grey-green surface. Adult **leaves** are lanceolate or falcate, 9–14cm long and to 3cm wide. The club-shaped **buds** have a conical cap shorter than the base. White to cream **flowers** in axillary clusters of 7–15, often in pairs, are produced in summer and autumn. Pear-shaped **fruits**, less than 1cm in diameter, have level or slightly enclosed valves.

This tree is easily killed by fire and, unlike other eucalypts, does not have lignotubers or epicormic buds to regenerate. Although fire penetrates Mountain Ash forests only once every 100 years or so, when it does, the blaze is intense. This species depends on fire to open the seed capsules and the abundant nutrients in the cleared ash bed favour seedling growth. As a result, Mountain Ash forests tend to be of even age and height. The pale brown timber is used widely for interiors and building construction. It is also considered the most important eucalypt by the pulp and paper industry.

Red Ironbark *Eucalyptus sideroxylon* 10–35m

This beautiful tree can be seen in dry woodlands and open forests from south-eastern Queensland over a wide area of inland NSW and into north-eastern Victoria. It is noted for its magnificent, deeply furrowed, dark brown to almost black **bark**, which persists to the small branches.

Drooping adult **leaves** are lanceolate to narrow-lanceolate, 7–14cm long and 1.8cm wide, and greyish-green on both sides. Ovoid **buds** to 1.2cm long have a conical cap shorter than the base. **Flowers** are produced in axillary clusters of 7 on long, slender stalks through winter and early spring. They are often very showy and come in various shades of white, cream, pink or red. Also on long stalks are the barrel-shaped **fruits**, about 1cm long with enclosed valves. Often cultivated because of its beautiful bark, foliage and flowers, the red ironbark is extremely adaptable in warm temperate to semi-arid areas. It will tolerate smog, hot summers, dry spells and frost.

A very similar ironbark, *E. tricarpa*, was formerly known as *E. sideroxylon* subsp. *tricarpa*. It differs mainly from *E. sideroxlyon* in having buds, flowers and fruits in clusters of 3. *E. tricarpa* occurs on the south coast of NSW and is common in Gippsland and central Victoria. Both species produce very hard, strong, durable dark red timber used for heavy construction and railway sleepers. The nectar-rich flowers are attractive to a variety of birds and are an important source of honey.

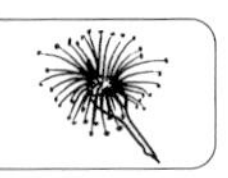

Silvertop Ash *Eucalyptus sieberi* 15–35m

Extensively distributed from the central coast and ranges of NSW, across eastern Victoria, usually south of the Great Dividing Range, Silvertop Ash also has an isolated occurrence north-west of Melbourne as well as in north-eastern Tasmania. Depending on location, it ranges in height and form from an impressive tall straight tree in eastern Victoria to somewhat stunted specimens on poorer soils in NSW.

The deeply fissured, grey-brown **bark** is persistent on trunk and larger branches with the upper parts a contrasting smooth grey or white and shedding in long ribbons. The bark on young trees is noticeably flaky and is orange-brown when fresh. The curved adult **leaves** are lanceolate, 9–15cm long and to 2.8cm wide, and glossy green on both sides with conspicuous lateral veins. Gum tips in summer are an attractive shiny red. Small, club-shaped **buds** have a short rounded cap with a small point. Profuse white flowers are borne in umbels of 7–15 in spring and summer. Conical or pear-shaped **fruits**, about 9mm across, have a broad, reddish disc and 3–4 slightly enclosed valves.

The light brown to pinkish timber is valued for construction, flooring, railway sleepers, poles, fencing, plywood and pulp. Silvertop Ash is the main species used for woodchipping.

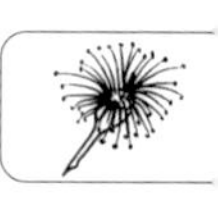

Darwin Stringybark *Eucalyptus tetrodonta* 15–25m

This is one of the dominant eucalypts in tropical open forests and woodlands extending across northern Australia from the Kimberley region, WA, the NT and in Cape York Peninsula, Queensland. It often grows in association with Darwin Woollybutt, *E. miniata*.

It is usually an erect medium-size to tall tree with an open crown and pendulous foliage. The rough, grey, fibrous **bark** is stringy and persistent on the trunk and branches. The curved lanceolate adult **leaves**, 10–25cm long and 1.5–5cm wide, are dull green on both surfaces and have an elongated pointed tip. Prominently ribbed club-shaped **buds** have 4 small teeth on the calyx and a domed cap longer than the base. Creamy white **flowers**, about 2cm across, are borne in umbels of 3 during winter to early spring. The bell-shaped, slightly ribbed **fruits**, to 2cm long and 1.2cm wide, have 4 prominent teeth around the rim and enclosed valves.

The pale red, fairly hard wood is used for posts and poles and general construction in the NT. Aborigines use the wood to make spears, woomeras, paddles and axe-handles and the bark is used for constructing shelters, canoes, utensils and paintings. The inner bark and leaves have many medicinal uses, including the treatment of sores, oral infections and to alleviate the pain of aching bones and muscles.

Lilly Pilly *Acmena smithii* 8–30m

Famous for its highly decorative berries, the Lilly Pilly occurs naturally over a large range from Gippsland, eastern Victoria to Cape York Peninsula, Queensland. It is widespread from the coast to the ranges, usually in rainforests and in sheltered places along watercourses. It reaches its maximum height in sheltered rainforests and may be reduced to a low, compact shrub close to the sea.

The **bark** is reddish-brown and scaly on the lower trunk and smooth on the upper trunk and branches. The dark green **leaves** vary in shape from narrow-lanceolate to broad-ovate or elliptic, mostly 3–11cm long and 1.5cm wide, with a long blunt tip. The upper leaf surface is glossy with a paler underside. New growth is an attractive bronzy pink. A smaller-leaved, lower growing form, var. *minor*, which occurs from about Newcastle to Bunya Mountains, has broad-ovate leaves 1–4cm long and only about 2mm wide.

The creamy white or pinkish **flowers** have 4 petals, 4 sepals and numerous stamens. They are small but borne in showy terminal panicles mainly during summer. The edible **fruit** is a rounded succulent berry, about 2cm across, white, pale mauve or purple hanging in great showy clusters when ripe in late autumn and early winter.

This handsome long-lived tree is an ideal specimen and shade tree for parks and large gardens. It will withstand close trimming and may be clipped to form an attractive bushy hedge or screen.

Lemon Ironwood *Backhousia citriodora* 3–20m

The beautiful lemon-scented foliage of this bushy shrub to medium-size tree yields a valuable commerical oil that is more than 90 per cent citral. Lemon Ironwood can be seen naturally in subtropical rainforests along a coastal strip between Mackay and Brisbane, Queensland. It is also popular in cultivation, especially in Queensland.

The shiny green narrow-elliptic **leaves**, 5–12cm long and 2.5cm wide, which taper to a blunt tip, have numerous oil dots and are pale and softly hairy on the undersides. Masses of creamy white **flowers** with long stamens are borne in umbel-like arrangements near the ends of branchlets through the summer months. The **fruit** is a small dry capsule with persistent calyx lobes.

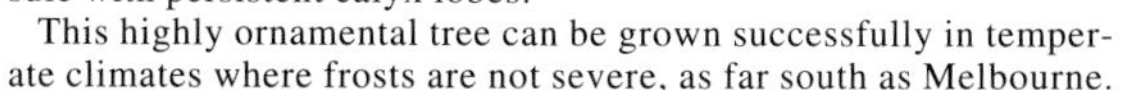

This highly ornamental tree can be grown successfully in temperate climates where frosts are not severe, as far south as Melbourne. The blossoms emit a delightful evening fragrance and are attractive to honeyeaters.

There are 8 species of *Backhousia*, all endemic to Australia. Most occur in rainforests of eastern Australia and most are known for their strongly aromatic foliage. A rare but interesting species, *B. anisata*, confined to the Bellinger and Nambucca valleys in north-eastern NSW, has strongly aniseed-scented leaves.

Willow Bottlebrush *Callistemon salignus* 3–10m

This widespread bottlebrush grows freely in moist locations, such as along the banks of streams, north from the Victorian border to south-eastern Queensland. It is a weeping tall shrub to medium-size tree with beautiful white papery **bark** and bright pink silky young leaves. An alternate common name is Pink Tips. The mature **leaves** are narrow-elliptic, up to 10cm long, with distinct lateral veins and numerous oil dots. Creamy white **flowers** are borne in bottlebrush spikes up to 5cm long and 3.5 across in spring. The **fruit** is a small rounded capsule about 5mm in diameter.

Red, pink or mauve flowering forms make this is a popular plant in cultivation.

Weeping Bottlebrush *Callistemon viminalis* 2–8m

This species grows as a shrub, small bush or tree. It is common along river banks in coastal stretches from northern Queensland to northern NSW. The **bark** is dark and furrowed. Linear **leaves**, slightly narrower than those of the willow bottlebrush and to 7cm long, are covered with aromatic oil glands. New-growth foliage is often bronze. Spikes of bright red **flowers**, from 4–10cm in length, are borne mostly in spring, but are seen intermittently at other times. The flowers are widely spaced and can be identified by the stamens, which are joined into tiny rings at their bases. The **fruit**, a small cup-shaped capsule, sheds the ripe seeds.

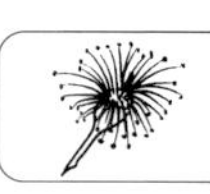

Brush Box *Lophostemon confertus* 35–40m

This handsome tree is widespread in rainforests of eastern Australia from coastal lowlands to mountains, from the NSW central coast, to Cooktown, Queensland. It can also be seen on the margins of rainforests and in wet sclerophyll forests. It has a straight trunk 2m or more in diameter in large specimens, and a dense crown of tiered dark green foliage.

The **bark**, rough and fibrous at the base, is shed on the upper trunk and branches, leaving an attractive, smooth pinkish-brown surface. The leathery elliptical to ovate-lanceolate **leaves**, 9–17cm long and 4.5cm wide, appear to be in false whorls. Undersides of leaves and young growth are covered with fine hairs. A sticky, whitish resin exudes from cut branchlets and leaf stalks. Axillary clusters of 7 white **flowers**, each about 2cm across, with 5 circular petals and numerous feathery stamens, are borne in summer and attract honeyeaters. The **fruit** is a 3-valved, bell-shaped capsule, about 1.5cm in diameter, with numerous fine seeds.

Brush Box, formerly known as *Tristania conferta*, is highly regarded as a timber tree, providing hard-wearing and attractive pinkish-brown timber with a wide variety of uses. It is a popular cabinet timber and was used for the Sydney Opera house floors. It is one of the hardiest and most planted Australian trees in streets, parks and as shelter for stock. Well-grown specimens make excellent shade trees.

Silver-leaved Paperbark *Melaleuca argentea* 10–20m

This small to medium-size tree is common across tropical northern Australia, where it favours sandy banks of freshwater waterways. It usually has pendulous branches and creamy white to soft, papery, grey **bark**. The mature **leaves** are grey green, narrow lanceolate, 6–12cm long and 6–12mm wide, with 3–5 main longitudinal veins and a pointed tip. *Argentea* means 'silvery' and refers to the young, silky hairy, silvery leaves. The greenish to cream **flowers** are borne in loose cylindrical spikes, 5–12cm long and about 3cm wide, in winter and early spring. Small, woody, cup-shaped **fruits**, about 4mm wide, contain many very fine seeds. The Aborigines use the papery bark for shelters, bedding and carrying items and an infusion of the leaves is used to treat colds and aches.

Cajuput Tree *Melaleuca cajuputi* 1–30m

This tall, straight, tropical tree occurs in northern Australia from eastern Kimberley, WA, across the Top End of NT to Cape York Peninsula, Queensland, where it grows around seasonal swamps, coastal plains and stream banks. Its range also includes Indonesia, Malaysia and South-East Asia. It has layered grey to white papery **bark** and stiff, narrow-elliptic dull green **leaves**, 5–10cm long and 1–2cm wide, with 5–7 longitudinal veins. White **flowers** are borne in dense cylindrical spikes 4–8cm long and about 2.5cm wide, usually in groups from late autumn to early spring. Small cup-shaped **fruits**, about 4mm wide, are clustered along the stem. Aborigines use the bark for many purposes and an infusion of leaves as a decongestant.

Saltwater Paperbark *Melaleuca cuticularis* to 10m

This shrub or small tree with rigid, twisted branches and soft layers of white papery **bark** is very common in south-west WA, where it occurs mostly in saline depressions, along estuaries or bordering lakes. The opposite, linear-oblong **leaves,** about 6–12mm long, are thick, flat or concave with a blunt tip. Profuse, creamy white **flowers** are borne singly or in small clusters of 2–3 at the ends of branches during spring. The **fruit** is a bell-shaped capsule, about 1cm in diameter, with 5 persistent out-turned lobes. The opposite leaves and unusual fruit separates this species from other tree-size melaleucas in south-west WA.

Swamp Paperbark *Melaleuca ericifolia* to 12m

A common species of swampy situations and along river banks, this paperbark is distributed mainly in coastal districts south from northern NSW to Tasmania. It is a large bushy shrub or small tree that frequently forms thickets by its habit of spreading suckers. The grey **bark** is papery. Dark green linear **leaves** are 1.5cm long and about 1mm wide. Creamy white scented **flowers** are borne in dense terminal spikes to 2.5cm long in late spring and summer. The **fruit** is a cylindrical capsule to 4mm across.

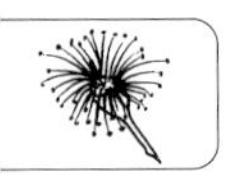

Moonah *Melaleuca lanceolata* 6–8m

Also known as Black Tea Tree, this small tree is fairly widespread in drier areas across the southern part of Australia from the eastern edge of the Nullarbor Plain, SA, to the central west of NSW and the Darling Downs district of southern Queensland. It has hard, grey to blackish corky **bark** and a very dark green rounded bushy crown. The linear to narrow-elliptic **leaves**, to 1.5cm long and 3mm wide, have a pointed tip. White or cream **flowers** are borne in short spikes, about 4cm long, mainly in summer. The barrel-shaped **fruits**, about 5mm in diameter, have a small opening.

Weeping Paperbark *Melaleuca leucadendra* 10–30m

Common across tropical northern Australia from the Kimberley region, WA, to Cape York Peninsula, and south to Bundaberg, Queensland. It occurs along watercourses and sometimes around swamps and can grow into a large tree with pendulous branches. The papery **bark** is whitish to pale brown. The dark green thin-textured lanceolate **leaves**, to 20cm long and 3cm wide, have 5–6 longitudinal veins. Prolific creamy white **flowers**, often in groups of 3, are borne in spikes about 15cm long, mainly in autumn and winter. The cup-shaped **fruits** are about 4mm in diameter. Aborigines infuse the leaves to ease headaches and the symptoms of coughs and colds. The papery bark has many uses.

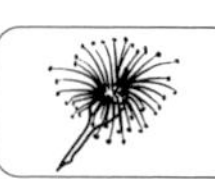

Narrow-leaved Paperbark *Melaleuca linariifolia* to 10m

This paperbark inhabits the edges of swamps or creek beds along the coast and adjacent ranges from south-eastern Queensland to southern NSW, as well as an isolated occurrence in central Queensland. It is a tall shrub or small tree with slender branches and cream-coloured papery **bark**. The dark green linear-lanceolate **leaves**, about 4.5cm long and 3mm wide, are mostly opposite. In early summer, the plant is covered with masses of creamy white **flowers** in spikes to 4cm long, which gives this species the alternate common name of Snow-in-summer. The flowers are attractive to a variety of nectar-feeding birds and insects. Small cylindrical **fruits** are about 4mm in diameter.

Broad-leaved Paperbark *Melaleuca quinquenervia* 10–15m

This common eastern species has a wide distribution along the coast from northern Queensland to southern NSW, where it frequents swampy sites often forming pure stands. It is a small to medium-size tree with attractive thick cream to grey papery **bark**. The thick-textured lanceolate to oblanceolate **leaves**, to 7cm long and 2.5cm wide, have 5 prominent longitudinal veins. Creamy white **flowers** are borne in spikes to 5cm long, mainly in summer and autumn. The woody cup-shaped **fruits** are about 5mm in diameter. This is a particularly useful and ornamental tree for growing in poorly drained moist sites.

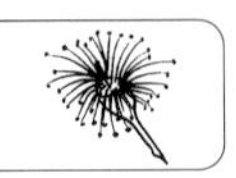

Turpentine *Syncarpia glomulifera* 40–45m

Turpentine is a tall, shaft-like tree found in wet sclerophyll forests chiefly in coastal districts from near Batemans Bay, NSW, to near Atherton in northern Queensland. It is highly regarded for its strong durable timber, which is extremely resistant to decay and marine borers and widely used for wharf construction and decking. It is mostly encountered in mixed forests on the margins of subtropical rainforests and may be associated with Tallowwood, *E. microcorys*, Flooded Gum, *E. grandis*, and Brush Box, *Lophostemon confertus*.

The tree has a stringy fibrous **bark** persistent over the trunk and branches. The stiff elliptic to ovate **leaves**, 7–11cm long and 4.5cm wide, are dull green on the upper surface and whitish-grey below. They are in opposite pairs and encircle the stems in groups of 4. Small rounded clusters of creamy white **flowers** with numerous stamens are borne on long stalks in spring and summer. The characteristic **fruit** is a hard woody capsule composed of 7 fruits fused into a rounded head about 2cm across, and containing numerous tiny seeds.

The name Turpentine refers to the tree's orange-red resinous exudate, which is said to resemble the highly flammable turpentine derived from certain species of pine (*Pinus*). As the timber of the Turpentine is one of the most resistant to damage by fire in the world, this common name is misleading.

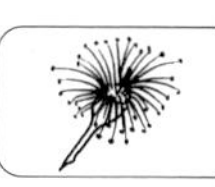

Water Gum *Tristaniopsis laurina* to 30m

Favouring shaded coastal gullies, Water Gum is mostly found in rainforests, usually besides streams. It is quite common and has a wide distribution from eastern Victoria to south-eastern Queenland.

This tree of variable size usually has a dense, dark green spreading canopy. The pale brown **bark** is rather smooth, but can have a rough appearance owing to the shedding of ribbony strips. The alternate leathery **leaves** are narrow-oblanceolate, 6–12cm long and to 3cm wide, shortly pointed and shiny dark green on the upper surface with a greyish-white undersurface. Scattered oil dots may be visible. Axillary clusters of 3 bright-yellow flowers about 1cm across, each with 5 oval petals, are produced during the summer months. The **fruit**, a 3-valved capsule 1cm long and 6mm in diameter, splits in winter when ripe.

Because of its ornamental appearance and relatively slow growth, Water Gum, formerly known as *Tristania laurina*, is commonly planted in eastern Australia, where it takes on smaller proportions and develops into a compact bushy tree. The wood is strong and hard and was used to make carpenters' tools, such as mallets, planes and tool handles.

The Mountain Water Gum, *Tristaniopsis collina*, has previously been confused with *T. laurina*, but differs in having persistent rather than smooth bark, and leaves drawn out to a narrow point with more numerous oil glands. It favours higher altitudes and is rarely found beside streams.

Crow's Ash *Flindersia australis* 15–40m

This species produces one of the hardest and most durable of Australian timbers. Used widely for general building purposes, the wood is yellow-brown, close-grained and greasy and resembles Indian Teak. An alternate common name is Australian Teak.

It is a large tree of dry and subtropical rainforests of north-eastern NSW to around Mackay, Queensland. The trunk is tall and straight and the scaly, mottled grey **bark** is shed in rounded flakes. Large trees may be slightly buttressed. Pinnate **leaves** 5–35cm long comprise 5–9 elliptic to narrow-ovate leaflets up to 12cm long and 4cm wide. Both surfaces are glossy green with prominent oil dots. It is semi-deciduous and usually sheds leaves from branches that are about to flower.

The small white to cream **flowers** are borne in dense terminal panicles, up to 15cm long, in spring. The distinctive brown **fruit** is a woody capsule, about 10cm long, consisting of 5 boat-shaped valves that remain united at the base and open out into a pointed star. The outside of the capsule is covered with short, blunt prickles and the seeds are prominently winged. The fruit is highly decorative and is much used in dried floral work and craftwork.

Crow's Ash, a popular ornamental shade tree, develops a shapely, dense, rounded crown and is much shorter in stature when grown in open situations. The foliage is used as a dyeing material.

Leopard Wood *Flindersia maculosa* 10–15m

The common name refers to the **bark**, which is shed in patches, giving a conspicuous mottled white, cream, brown and grey 'leopard-skin' appearance to this distinctive small tree of inland NSW and central Queensland. It grows in open, dry situations on sandplains and low-lying areas subject to some seasonal flooding and sometimes on stony hills. Mature trees have a straight single trunk and a well-formed spreading rounded crown. Juvenile plants start as a much-branched spiny shrub (see bottom photo). A single stem from this shrub eventually becomes erect and forms the primary trunk.

The dark green shiny **leaves** are narrow-elliptic to oblong, 3–7cm long and to 1cm wide, and have a light green underside. Small cream tubular **flowers** are borne in loose terminal panicles, to 7.5cm long, in late spring. The **fruit** is a woody, 5-valved, dark brown capsule, 2–3cm long, covered with short hard prickles. Each valve separates fully and sheds several flat seeds, which are about 2cm long and winged at both ends. The foliage of mature trees provides useful stock fodder in times of drought. Those from the juvenile form (bottom photo) are apparently unpalatable.

Wilga *Geijera parviflora* 4–9m

This tree, often seen in rural areas, looks as if it has been neatly trimmed beneath. Wilga is loved by stock, especially sheep, which constantly graze on the lower leaves. In times of drought, plants are often cut and fed to stock. Wilga is widespread in semi-arid regions of Queensland, NSW, north-western Victoria and northern SA. It is a small spreading tree with a short upright trunk with grey **bark**. The trees develop a well-shaped rounded crown of pendulous branches that almost sweep the ground. The linear-lanceolate pendent **leaves**, 3–18cm long and to 1cm wide, are shiny, dotted with oil glands and strongly aromatic when crushed. Aborigines use the chewed leaves to pack into tooth cavities to stop toothache. Very small creamy white tubular **flowers** with a strong scent are borne in loose terminal panicles, 4–7cm long, in winter and spring. The rounded **fruit**, about 5mm in diameter, has thick outer skin and contains a single glossy black seed.

This is an excellent shade and shelter tree, highly recommended for planting throughout rural areas. A deep, well-drained soil and a sunny open position with plenty of room to spread is ideal. It resists drought after the first year and will tolerate frost.

Pink-flowered Doughwood *Melicope elleryana* 12–20m

This is a fairly common tree in coastal subtropical rainforests and swamp forests north from Yamba, NSW, to northern Queensland and extending across the tropics to the north-eastern Kimberley region, WA. Formerly known as *Euodia elleryana*, it is a small to medium-size tree with a slender trunk and smooth light grey or brownish corky **bark**. It has a light widely spreading canopy and opposite trifoliate **leaves**. The elliptic leaflets are soft, 7–20cm long and 3–9cm wide, with numerous small oil dots. The foliage is mostly clustered toward the ends of the branches. Many small pink **flowers** are borne in compact dense clusters in the leaf axils mostly in summer and early autumn. Occasionally, white flowers are found. The **fruit** is a small capsule made up of 4 almost separate lobes.

This species has been introduced to cultivation and is a popular fast-growing ornamental and shade tree in subtropical regions. Flowers and berries are both attractive to native birds. In north-eastern Australia, the leaves are eaten by the caterpillars of the beautiful Ulysses Butterfly, noted for its vibrant electric blue and glossy black wings.

Rosewood *Dysoxylum fraseranum* 2–40m

This common rainforest tree of the coastal ranges from south-eastern Queensland to Wyong, NSW, favours subtropical rainforest, where it grows tall and slender with a spreading crown and light brown, scaly **bark**. The pinnate **leaves**, to 25cm long, are composed of 4–12 elliptic or obovate leaflets, each about 11cm long and 4cm wide, with entire margins and raised glands on the undersides. The small, fragrant, creamy white **flowers**, about 1cm across, are borne in panicles in the upper leaf axils in autumn and early winter. The **fruit**, a rounded or pear-shaped capsule about 4cm in diameter, is rosy-red when ripe and splits open to reveal 6–8 shiny red seeds. The reddish-brown timber has a rose-like perfume and is used for furniture, carving and internal fittings.

Red Bean *Dysoxylum mollissimum* 15–25m

This medium to large tree, formerly called *D. muelleri*, is widespread in the subtropical rainforests north from the Bellinger River, NSW, to Cairns in northern Queensland. The grey **bark** is scaly. The large pinnate **leaves**, to 45cm long, have 11–23 soft ovate leaflets, each 3–15cm long and 2–5cm wide, with small tufts of hair present in most vein angles. Numerous cream **flowers**, to 1.5cm across, are borne in panicles up to 35cm long in the upper leaf axils in summer and autumn. The **fruit**, a yellowish-brown rounded capsule, about 2cm in diameter, has a wrinkled surface and opens in 3–5 valves. The easy-to-work reddish-brown timber is used for cabinet work and for indoor fittings.

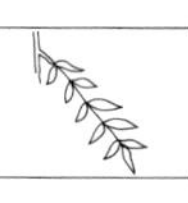

White Cedar *Melia azedarach* to 10m

Widely planted as a hardy ornamental tree, White Cedar is found naturally over a wide area, usually bordering rainforests in coastal districts from northern Queensland to as far south as the Shoalhaven River, NSW. It is also found throughout much of the Kimberley region, WA.

It is a small to medium-size deciduous tree with dark grey furrowed **bark**. The large **leaves**, to 45cm long, are bipinnate with pointed ovate leaflets, each about 10cm long and 3cm wide and usually variously toothed. The foliage is quite noticeable in autumn when it turns bright yellow before falling. Sweetly scented white to lilac **flowers**, with 5 spreading petals, are borne in loosely arranged panicles about 20cm long in late spring. The **fruit** is a rounded yellowish berry, about 1.5cm across. The fruits persist throughout winter and are reputed to be poisonous to humans and livestock, but are relished by a variety of native birds, including parrots.

Cultivated forms of White Cedar have been widely grown for many years throughout most parts of Australia and have become naturalised in many areas often making it difficult to know if some plants are native to a particular district. It is a beautiful perfumed shade tree, suitable for inland planting where it will tolerate most frosts and dry periods.

Red Cedar *Toona ciliata* to 40m

In the past, the famous Red Cedar was very much sought after for its beautiful reddish-brown timber. It had a wide distribution in coastal or near-coastal rainforests from northern Queensland south to Ulladulla, NSW, but because most accessible trees have been removed, few large specimens remain in the wild.

It is a tall deciduous tree with an open spreading crown. Some older trees may be buttressed and the greyish-brown **bark** is rough and scaly. The glossy pinnate **leaves**, to 45cm long, are composed of 4–10 pairs of ovate leaflets, each up to 15cm long and 5cm wide, with entire margins. In winter, the leaves turn brown and fall and new foliage is pinkish-red. Many small white or pale pink **flowers** are borne on pendulous panicles, 20–40cm long, in late spring. The **fruit** is a thin-textured, 5-valved oblong capsule about 2cm long and 8mm across.

One of the finest of Australian cabinet timbers, Red Cedar, formerly known as *T. australis*, was also once used for doors, window frames and staircases. Attempts to grow it as a plantation tree have been unsuccessful as the larvae of the cedar tip moth attacks the growing tips, preventing the young plant from developing into its normal tree form.

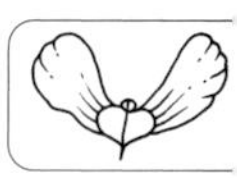

Western Rosewood *Alectryon oleifolius* to 7m

This small tree occurs in all mainland States and is widespread in semi-arid areas, often in open woodland on sandy or gravelly soil. It has silky new growth and the branches and leaves are often pendulous. The dark grey or brown **bark** is hard and lightly furrowed. Pale, dull green to greyish-green, narrow-elliptic **leaves**, 3–15cm long and to 2cm wide, are covered with very short hairs and have entire margins. The small pale green **flowers** have no petals and are borne in short racemes, to 9cm long, mainly in winter. The green **fruits**, about 1.5cm across, have 2 rounded segments each containing 1 glossy black seed with a fleshy red aril which, on some trees, is conspicuous.

Western Rosewood is an important stock fodder during periods of drought and healthy animals not subject to stress usually eat considerable quantities of the foliage without ill effect. Large amounts of young leaves, especially when wet, have been known to poison sheep and cattle when hungry animals have been travelling or are under stress. The attractive pinkish-brown timber is very hard and heavy.

The specific name refers to the similarity of the leaves to those of the olive tree, *Olea*. It was formerly known as *Heterodendrum oleifolium* and by many marvellous local common names, such as Cattle Bush, Bullock Bush, Boonaree, Boonareet, Minga, Jiggo, Behriging and Red Heart.

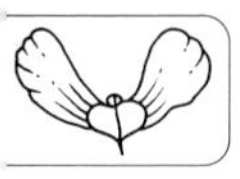

Whitewood *Atalaya hemiglauca* 7–15m

The Whitewood is found through most of dry inland Australia in all mainland States except Victoria. It is a very common tree and is sometimes the dominant species in woodlands of semi-arid areas. It is small and erect with a fairly dense crown and flaky **bark** that is pale grey or creamy coloured on the trunk and whitish-grey on the branches. The grey-green pinnate **leaves**, 8–20cm long, have 2–6 well-spaced oblong leaflets, each about 5–15cm long and 1.5cm wide. Masses of creamy white bell-shaped **flowers** with 5 spreading petals are borne in large panicles, about 20cm long, at the ends of branches in summer. The slightly hairy winged **fruit** is yellowish-green and about 4cm long, including the papery wing. When in flower and fruit, the ends of the branches become pendulous with the weight.

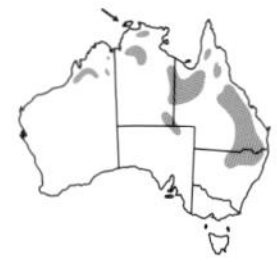

In inland regions, Whitewood is recognised as an important shade and shelter tree and the foliage is considered good fodder for stock. There is some evidence of toxicity at times, particularly in horses. The young shoots and the fruits are the most toxic part, especially when consumed in large quantities by very hungry animals when other feed is scarce.

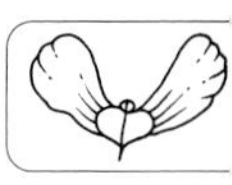

Tuckeroo *Cupaniopsis anacardioides* 8–15m

Common around Sydney, this tree extends along the east coast of NSW to northern Queensland and the Top End of the NT. It inhabits a variety of situations from windy headlands, sand dunes, along estuaries and littoral rainforests. It is a small to medium-size spreading tree in protected situations in rainforests, but is smaller in exposed coastal situations where it will remain a short dense shrub. The dark green pinnate **leaves**, to 25cm long, comprise 4–11 obovate to oblong leaflets that are prominently veined and notched at the tip. Flushes of new growth are often deep pink. Small greenish-white fragrant **flowers** are borne in axillary panicles, to 30cm long, in autumn and winter. The **fruit** is a 3-valved orange capsule, 3cm long and 2cm wide, that splits to reveal 3 black seeds, each enclosed in an orange covering (aril). The fruits ripen in summer and are most attractive to currawongs and other fruit-eating birds.

Tuckeroo makes an excellent shade tree, especially in warm coastal gardens, where it will tolerate considerable exposure to salt-laden winds. Plenty of water and regular applications of fertiliser will encourage steady growth.

Mulga *Acacia aneura* to 10m

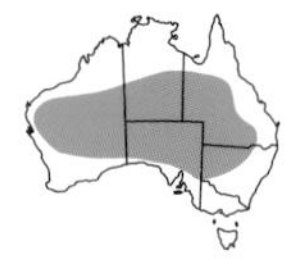

This shrubby tree is widely distributed over arid inland Australia in all mainland States, except Victoria. It often forms dense, extensive stands in tall shrublands or scattered in low woodland formations in association with saltbushes or tussock grasses. It has a short trunk, with dark grey fissured **bark**, and often branches from about 1m above ground, forming a spreading umbrella-like crown with upward pointed **phyllodes**. The grey-green phyllodes are variable in shape from almost cylindrical to narrow lanceolate, from 4–10cm long and up to 1cm wide, with a hardened point at the apex.

Golden yellow, cylindrical **flowerheads**, to 2cm long, are borne singly or, rarely, in pairs in the axils of the phyllodes. They occur at irregular periods throughout the year, mainly in autumn and especially after good rains. The **fruit,** a flat oblong pod, 4cm long and 1.5cm wide, is covered with raised, net-like veins. The flat, oval, dark brown **seeds** are highly nutritious and are roasted and ground for damper by Aborigines of central Australia. Green insect galls, about the size of a cherry, found on the foliage and known as Mulga 'apples', are also eaten in desert areas.

Mulga is an Aboriginal name for a long narrow shield made from the dark brown, yellow-grained wood, which is now used mainly for souvenirs and ornamental woodwork.

Coast Myall *Acacia binervia* 15–20m

This handsome erect spreading tree has an attractive greyish green appearance. It was previously known as *Acacia glaucescens*, perhaps a more suitable name, a reference to its bluish-grey phyllodes. It is found in dry eucalypt forests in bushland round Sydney, in the lower Blue Mountains and in coastal areas of NSW, extending to the tablelands and central western slopes.

It has a thin, reddish-brown **bark** and its sickle-shaped **phylodes**, to 15cm long and 2cm wide with 2 or 3 prominent longitudinal veins, have a silvery grey sheen. Young foliage is often covered in soft hairs. In early spring, abundant yellow, cylindrical **flowerheads**, up to 6cm long, are produced singly or in loose clusters in the axils of the phyllodes. The **fruit,** a straight, flat pod to 8cm long and 5mm wide, is slightly raised over the seeds.

This popular, long-lived wattle makes an outstanding specimen tree for cultivation in streets, parks and large gardens. It is also suitable for windbreaks on rural properties, but prussic acid in the young foliage is reported to be poisonous to stock.

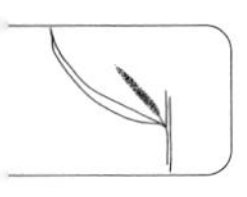

White Sally *Acacia floribunda* 4–8m

Widespread in open eucalypt bushland around Sydney, this species also occurs in coastal areas of southern Queensland and East Gippsland, Victoria. It is a tall shrub or small tree with somewhat pendulous branches and narrow, lance-shaped **phyllodes**, 5–15cm long and 2–10mm wide. These are thin in texture, have several fine parallel veins, with 2 to 4 being more prominent, and taper to a fine curved point. Glands are absent. Scented pale yellow cylindrical **flowerheads**, to 6cm long, are borne in axillary pairs in late winter and early spring. The **fruit** is a long, narrow pod, to 12cm long, ending in a long, thin point.

Sydney Golden Wattle or Sallow Wattle *Acacia longifolia* to 8m

Contrary to one of its common names, this widespread species is common along the coast and tablelands of NSW and extends into eastern Victoria. It is a small bushy tree with smooth, greyish **bark**. It is closely related to *Acacia floribunda*, but differs in having thicker and larger **phyllodes**, 6–20cm long and up to 2cm wide. These have 2, 3 or more prominent longitudinal veins and a single gland near the base. One or 2 golden-yellow cylindrical **flowerheads** to 6cm long are produced in the phyllode axils from late winter into spring. The **fruit** is a leathery, almost terete, pod to 12cm long and 6mm wide, often becoming twisted.

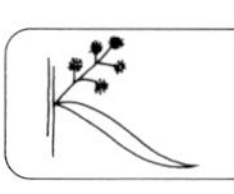

Ironwood *Acacia estrophiolata* 5–15m

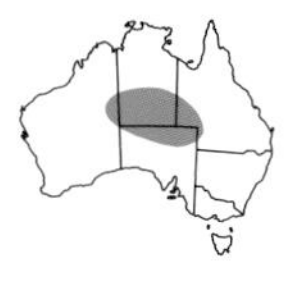

This interesting weeping tree occurs in arid central Australia and northern SA, where is grows mainly in open woodlands and shrublands, often on red soils. It has long drooping branches and dark grey to brown **bark**, which is rough and fissured, on the trunk and main branches. The linear grey-green **phyllodes** are straight or slightly curved, 4–10cm long and 2–4mm wide with 3 longitudinal veins. Pale yellow globular **flowers** are borne on slender stems singly or in pairs in the leaf axils mainly during winter or at irregular periods throughout the year. The **fruit** is a dark brown narrow pod, 3–10cm long and 7mm broad, unevenly constricted between the seeds.

The timber is very hard and durable and is used mainly for fencing. Aborigines use an infusion of the bark as a wash for the treatment of sores, boils and skin infections. This is a highly ornamental tree recommended for cultivation in hot inland areas. The name Ironwood is used for a number of different Australian trees, including *A. excelsa*, which also has drooping branches. It is widespread in western areas of Queensland and NSW and has ashy grey oblong leaves, 5–8cm long and to 2cm wide with 5–7 prominent veins.

Two-veined Hickory *Acacia binervata* 4–10m

This small tree ranges along the coast and tablelands of NSW, especially the Illawarra district, extending into mountainous regions of south-eastern Queensland. It prefers moist shelter in tall open forests and on margins of warm temperate and subtropical rainforests. It has smooth or finely fissured **bark** and densely hairy branchlets. The broad, lance-shaped **phyllodes**, 7–12cm long and up to 3cm wide, are often curved and usually have 2 (sometimes 3) prominent veins, which help to identify this species. A distinct gland occurs on the upper margin near the base of the phyllode. Pale yellow or cream globular **flowerheads** are produced in large axillary panicles during spring. The **fruit** is a flat, thin-walled, smooth pod, 7–14 long and up to 1.5cm wide.

Brisbane Wattle or Fringed Wattle *Acacia fimbriata* 5–8m

Spectacular when in flower, this bushy shrub or small tree grows in dry eucalypt forests and woodlands mainly in coastal districts of NSW and south-eastern Queensland. The slender branches often droop, especially when laden with blossom. The linear, dark green **phyllodes**, 2–5cm long and up to 6mm wide, are fringed with minute hairs. They have a pimple-like gland near the base and taper to a fine, non-pungent point. Although individual globular golden-yellow **flowerheads** are small, up to 20 are borne on axillary racemes extending beyond the foliage and almost covering the plant during September. The **fruit** is a flat, thin-textured pod, 5–8cm long and up to 6mm wide.

Brigalow *Acacia harpophylla* 10–20m

Found west of the Great Divide in parts of NSW as well as in extensive areas in Queensland, where it grows in large stands often tall and dense enough to form a low impenetrable forest known as Brigalow scrub. It has dark, almost black, furrowed **bark**. The silvery grey **phyllodes**, 10–20cm long and up to 2cm wide, are curved and taper equally to both ends. These have 3 to 5 rather prominent longitudinal veins and a raised gland at the base. The mid-yellow, globular **flowerheads** are produced in short axillary racemes. Flowering is somewhat irregular, usually in late winter and early spring. The **fruit** is an almost cylindrical pod, 3–11cm long and up to 1cm wide, with slight constriction between the seeds.

Club-leaf Wattle *Acacia hemignosta* 4–7m

This attractive small bushy tree is scattered across northern Australia extending from the Kimberley region, WA, and NT to the coast near Cairns, Queensland. It has pendulous branches and brown to grey corky **bark**. The oblong to obovate bluish to grey-green **phyllodes**, 7–10cm long and to 2.5cm wide, have 3 longitudinal veins and a rounded tip. Globular yellow **flowers** are borne mostly in axillary or terminal racemes, up to 12cm long, or occasionally singly in the leaf axils during winter. The **fruit** is an oblong, light brown papery pod, 5–7cm long and to 1cm wide, with a thickened margin.

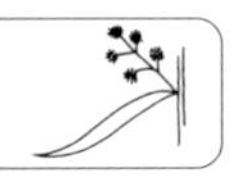

Blackwood *Acacia melanoxylon* 8–30m

One of the largest of the acacias, Blackwood is a tall symmetrical tree sometimes reaching a height of 30m or more in favourable situations. It is widespread in eastern Australia as far north as the Atherton Tablelands, Queensland, and in higher rainfall areas in south-eastern SA and Tasmania. It develops best in cooler southern areas, especially in western Tasmania, where it favours cool temperate rainforests.

The dark grey **bark** is hard and furrowed. The broadly, lanceolate dull green **phyllodes**, 6–14cm long and 1–3cm wide, have 3–5 prominent longitudinal veins and a small blunt point. Masses of pale yellow globular **flowerheads** appear on short branched racemes from late winter to early summer. The **fruit**, a rather flat pod 4–12cm long and up to 1cm wide, is often twisted or curved in a circle; orange to pinkish-red seed-stalks encircle the seeds in a double fold. Blackwood, so-called for the darkness of its mature wood, is strong and close-grained, often beautifully figured, and is considered one of the world's best furniture timbers.

Acacia implexa, Hickory Wattle, is closely related to Blackwood and grows in similar areas, but not in Tasmania or SA. Its phyllodes are somewhat shorter and the golden-yellow flowerheads are produced mostly in autumn. An easy way to distinguish this species from *A. melanoxylon* is by the seeds: the thin seed-stalk is paler and folded under the seed.

Boree or Weeping Myall *Acacia pendula* 6–12m

Widespread in low-rainfall areas of inland Queensland, NSW and Victoria, this beautiful weeping tree can often be seen as pure stands on heavy black fertile soils on major river floodplains. It has grey furrowed **bark** and pendulous branches, often drooping almost to the ground. The silvery green **phyllodes**, 5–14cm long and up to 1cm wide, are curved and have 1–3 longitudinal veins. They taper to a soft hooked point and have a single gland near the base. Small, pale yellow globular **flowerheads** are arranged in a short axillary raceme. Flowering is somewhat erratic, but mainly in summer and autumn. The **fruit** is a flat pod, 5–10cm long and about 2cm wide, with prominently winged margins.

The aromatic timber, often described as having the perfume of violets, is one of the hardest and heaviest of the acacias. It is beautifully grained and takes on a high polish; it is suitable for turnery and small ornamental articles. In drier parts of Australia, the foliage is used as stock fodder in times of drought.

Conspicuous brown silky bags are often seen hanging from the branches. These are used as shelter bags by day for procession caterpillars, which emerge at night and feed on the foliage, often causing considerable damage.

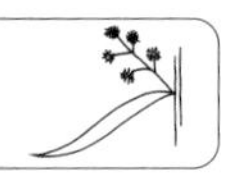

Golden Wattle *Acacia pycnantha* 3–8m

This tall shrub or small open-branched tree is Australia's official floral emblem and is featured on the coat-of-arms and currency. Native to SA, Victoria and south-western NSW, it is mostly found growing in open eucalypt forests on dry sandy soils or stony ground.

It has smooth, dark brown or grey **bark** and often pendulous branches. The thick and leathery bright green **phyllodes**, 6–20cm long and 1–4cm wide, are usually broadest above the centre and have a strong central vein and numerous conspicuous lateral veins. One or 2 glands are present on the upper margins. In late winter and spring, profuse, large and perfumed golden yellow globular **flowerheads** are produced on thickened stems in axillary racemes up to 15cm long. The **fruit** is a straight or slightly curved brown pod, 5–14cm long and up to 1cm wide, with margins slightly constricted between the seeds.

The bark is a well-known source of tannic acid and is considered one of the best tanbarks in the world. Nectar produced from the phyllode glands is relished by honeyeaters and small insectivorous birds. Golden Wattle is a highly recommended ornamental shade tree for areas that receive good winter rainfall.

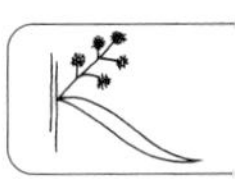

Golden Wreath Wattle *Acacia saligna* 3–10m

Native to the south-west of WA, this species is widely cultivated and, because of its ability to germinate freely, has become a weed in parts of Queensland, NSW, Victoria and SA, especially in near-coastal districts.

It is a dense bushy shrub or small tree, with smooth, grey to reddish-brown **bark** and often produces suckers. The variable green to glaucous drooping **phyllodes**, 8–30cm long and 6–50mm wide, straight or curved, have a prominent mid-vein and fine, but distinct lateral veins. They taper gradually toward the base and tip with one gland very close to the base. In late winter and early spring, bright golden yellow globular **flowerheads** are produced in extended axillary racemes. The **fruit** is a straight or slightly curved brown pod, 5–14cm long and about 5mm wide, with slight constriction between the seeds.

This species is sometimes confused with *Acacia pycnantha*, which differs in having stout zigzagging flowering stalks and a more prominently tapered asymmetric phyllode base. Golden Wreath Wattle was once the main source of tanbark in WA and has been extensively cultivated in South and East Africa and other countries for this purpose.

Cootamundra Wattle *Acacia baileyana* 3–7m

One of Australia's best-known wattles, this large spreading shrub or small dense tree is endemic to a limited area around Cootamundra, southern NSW. It has been widely planted and is now naturalised in most States. It has smooth, brown **bark** and glaucous branchlets. The silvery green bipinnate **leaves**, 2–4cm long, comprise 3–4 pairs of pinnae and 12–24 pairs of leaflets. These tiny leaflets are linear-oblong, 5–7mm long and about 1.5mm wide. Profuse, bright yellow globular **flowerheads** are borne in dense racemes from the leaf axils in winter and early spring. The fruit is a straight, flat pod, 4–10cm long and about 1cm wide. The seeds are relished by cockatoos and rosellas.

Silver Wattle *Acacia dealbata* to 25m

Although it reaches only about 15m in the NSW tablelands, 25m specimens occur in mountain forests in parts of Victoria and Tasmania. It has dark grey to almost black **bark**, smooth to deeply furrowed. Branchlets, foliage and young growth are silvery white, due to a covering of short white hairs. The bipinnate **leaves**, to 10cm long, comprise 10–24 pairs of pinnae and up to 45 pairs of leaflets. There is a conspicuous gland at the junction of each pair of pinnae. The sparsely pubescent leaflets are linear-oblong, 2–5mm long. Masses of pale lemon to bright yellow globular **flowerheads** are borne on extended axillary racemes to 10cm long in late winter and spring. The **fruit** is a light purplish-brown pod, 6–10cm long and about 1cm wide, with almost straight margins.

Black Wattle *Acacia decurrens* 5–15m

The specific name of this wattle refers to the prominent wings or ridges running down the stems from the base of the leaves. This characteristic helps to distinguish it from its nearest relatives *Acacia mearnsii* and *A. parramattensis*. An erect spreading tree found on the coast and tablelands of NSW, Black Wattle is common in the Sydney region. It has dark grey, furrowed **bark**. The deep green bipinnate **leaves** have 4–12 pairs of pinnae with up to 35 pairs of very narrow, widely spaced leaflets. The fragrant, brilliant yellow globular **flowerheads** are carried in fairly long axillary racemes in late winter and early spring. The **fruit** is a flat, linear, dark brown pod, 5–10cm long and 5–8mm wide.

Mountain Cedar Wattle *Acacia elata* 10–20 m

This species, one of the tallest of the wattles, can reach up to 20m high in gullies of rainforests and wet eucalypt forests of central and northern NSW. The **bark** is dark brown and deeply furrowed. Dark green, shiny bipinnate **leaves**, 15–22cm long, have 3–5 pairs of pinnae each with 10–20 pairs of leaflets. The individual leaflets are up to 6cm long and 1cm wide. The foliage resembles that of the pepper tree (*Schinus molle*). Axillary racemes of pale yellow globular **flowerheads** are borne in summer and early autumn. The **fruit** is a straight, brownish pod, 10–15cm long and about 1cm wide.

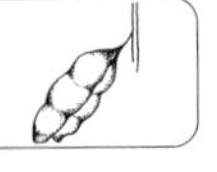

Northern Ironwood *Erythrophleum chlorostachys* 8–15m

Also known as Cooktown Ironwood, this distinctive semi-deciduous tree of northern Australia is widespread from north-eastern Queensland to the Kimberley region, WA, where it grows among eucalypts in tropical woodlands and open forests.

This species has a hard, coarsely tessellated, dark grey to black **bark** and a fairly dense canopy of spirally arranged bipinnate **leaves**. There are usually 5–9 dark green and leathery orbicular to ovate leaflets, 6.5cm long and 6cm wide. New leaves are bright green and appear just before the wet season. Yellow or greenish **flowers** are borne in dense spike-like racemes, 5–8cm long, in the upper leaf axils in late winter and spring. The **fruit** is a flat, dark brown to black pod, 10–20cm long and 1–4cm wide, containing 4–8 rounded, dark brown seeds.

All parts of the tree are extremely toxic to grazing animals and humans. The timber is very hard, dense and strong, termite-resistant and highly suitable for sleepers, fence posts and structural purposes. Aborigines use the wood for spears, axe-handles and digging, and for music and cooking sticks. A resin extracted from the roots is used as an adhesive for attaching spearheads to their shafts and in the construction of many items. In Aboriginal medicine, an infusion of the leaves and inner bark is used externally as a wash for sores, infected wounds and scabies.

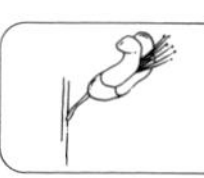

Black Bean *Castanospermum australe* 10–35m

The handsome Black Bean occurs in rainforests of eastern Australia north from around Lismore, NSW, to Cape York Peninsula, northern Queensland, where it is very common along the banks of streams and rivers. It has been cut extensively for its dark, attractively figured timber, highly valued by craftsmen for furniture, interiors and wood-carving.

This species is a tall tree with a straight trunk and a dark green, dense leafy crown. The **bark** is grey to brown and slightly rough. Pinnate **leaves** to 60cm long are made up of 9–17 oblong-elliptic leaflets, each up to 20cm long and 5cm wide. Showy, brilliant red to orange pea-shaped **flowers**, about 4cm long, are borne in racemes up to 15cm long, along the branches on old leafless wood in late spring. Flocks of honeyeaters and parrots, including lorikeets, are attracted to the nectar-rich flowers. The **fruits** are large pendent bean-like pods, 10–20cm long and up to 6cm across, with brown chestnut-like seeds that are poisonous.

Black Bean is an extremely popular shade tree and has been planted as a street tree for decades. When grown away from its rainforest habitat, it takes on smaller spreading proportions and rarely grows more than 10m high. Rich, well-drained soil and ample moisture are its main requirements. Once established, it will tolerate moderate frosts.

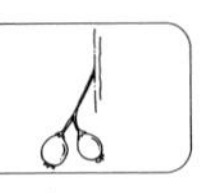

Cherry Ballart *Exocarpos cupressiformis* 4–8m

This small shapely tree has a very wide distribution over temperate eastern and southern Australia and is found in a variety of habitats, including various forests and woodland communities. It has a conifer-like appearance with slender drooping branchlets, often with bronze tonings. It has brown ridged **bark**. The **leaves** are reduced to minute scales. Tiny cream **flowers** are borne throughout the year in short spikes or clusters about 6mm long. Usually only one flower of the spike develops into a fruit.

The **fruit**, a globular nut about 5mm in diameter, is supported on a thickened flower stalk which is green at first, later becoming red and egg shaped. The fleshy stalk is edible and pleasant tasting and is one of the most widely known Aboriginal fruits in south-eastern Australia. The common name is derived from similar sounding Aboriginal names given to the fruit — Ballee, Ballat and Balad.

Species of *Exocarpos* are semi-parasitic on the roots of other plants so these attractive plants are usually difficult to establish in cultivation. The foliage is reputedly poisonous to stock.

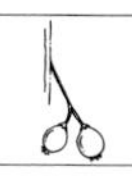

Quandong *Santalum acuminatum* to 6m

Best known of all Australian bush fruits, the Quandong has long been a favourite food of Aborigines. It was also used by early European settlers for jams and various desserts and these days can be bought as a delicious chutney or conserve in shops specialising in 'bush tucker' products.

A relatively common species, Quandong is widespread in inland areas of all mainland States. It is a large shrub or small tree with an open crown and spindly erect trunk. The tree is a root parasite and occurs in a range of mallee and woodland communities, as single specimens or in small stands. The light-brown **bark** is slightly furrowed. The lanceolate, pale olive-green **leaves**, 4–12cm long to 1.5cm wide, taper to a curved point. Numerous small whitish-cream **flowers** are borne sporadically throughout the year in short loose clusters at branch ends. The **fruit** is a fleshy, bright red globular drupe, 2.5cm in diameter, with a large, deeply pitted stone.When ripe, the fruit is slightly acid and can be eaten raw, but is more often stewed with sugar. The flesh can also be dried and stored for future use. The oil-rich kernel is pounded and used for cosmetic and medicinal purposes by Aborigines and the pitted stone is sometimes made into body ornaments and necklaces.

Laden with ripe fruit, Quandong is a most attractive plant and makes a good ornamental and shade tree for semi-arid gardens.

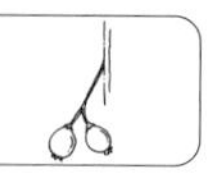

Northern Sandalwood *Santalum lanceolatum* 3–7m

Of about 25 species of *Santalum*, 6 are found in Australia. The Northern Sandalwood, or Plum Bush, is the most widespread, occurring across tropical and arid regions of northern Australia and inland NSW. An erect shrub or small tree with pendulous spreading branches, it has ovate to narrow-elliptic **leaves**, 2–9cm long and 5–25mm wide. These are bluish-green, thick and often slightly glaucous. The cream or pale green **flowers** are borne in panicles in the upper leaf axils mainly in spring and summer. The sweet, edible **fruit**, 1cm across, has a soft, dark blue or purple outer layer, a small smooth stone and a circular scar near the end. Although the timber is aromatic, this is not the commercial Sandalwood, *S. spicatum*, once extensively harvested in dry inland areas of WA and SA and exported to Asia.

Bitter Quandong *Santalum murrayanum* to 4m

Closely related to the common Quandong, this species has distinctive **leaves**, being opposite or often in groups of 3, narrow-lanceolate with short hooked tips and grey to yellowish-green in colour. It is a tall shrub or small tree, with pendulous branchlets, found in mallee communities across southern Australia. The rounded **fruit**, 2.5cm in diameter, is green to brownish-red when ripe. The outer layer does not separate easily from the stone and is inedible and very bitter.

Native Frangipani *Hymenosporum flavum* 4–20m

Widely cultivated for its highly fragrant flowers, the Native Frangipani naturally occurs over a wide range in eastern Australia north from around Sydney to north-eastern Queensland and New Guinea. This delightful small to medium-size tree grows in or near rainforests on fertile soils, reaching its best development at the northern end of its distribution.

It has a brownish-grey smooth **bark** and layers of spreading or sometimes horizontal branches. The soft-textured, glossy dark green **leaves** are lanceolate or obovate, 8–16cm long and 4.5cm wide, and taper to a blunt point. They are often in whorl-like clusters near the ends of the branchlets.

Loose terminal clusters of **flowers** are produced in great quantities in spring. Each tubular flower is about 5cm in diameter with 5 spreading lobes. They open a pale cream and change to a rich butterscotch shade with age. Some flowers have a purplish to deep tan centre. They are very sweetly perfumed, the fragrance being reminiscent of the exotic frangipani. The **fruit** is a brown ovate capsule, 2–3.5cm long, which splits and releases numerous flat, prominently winged seeds.

Native Frangipani is a beautiful, small specimen tree and is extremely reliable in cultivation, provided young plants are protected from frost and regularly watered. Once established, it is moderately frost-tolerant and has been successfully grown as far south as Melbourne.

Weeping Pittosporum *Pittosporum phylliraeoides* 3–10m

This widespread and interesting pittosporum occurs mainly in the drier areas of all mainland States. It is an important medicinal plant for Aborigines of the inland, who make a rubbing paste from the ground seeds for the relief of sprained limbs, cramps and skin irritations. An infusion of the leaves is taken for the relief of internal pains, and a compress of warmed leaves is applied to induce milk flow in new mothers. The tree also yields a good edible gum.

The pendulous branchlets give a graceful, somewhat willow-like appearance to this small tree. It is smooth throughout, except for the very young shoots, which may have greyish-brown hairs. The **bark** is whitish or mottled. The alternate **leaves** are oblong or linear-lanceolate, 4–12cm long and about 1cm wide, with a fine hooked tip.

Small, perfumed, cream or yellow tubular **flowers**, about 1cm long, have 5 spreading lobes and are borne singly or in small clusters from late winter to mid-spring. The **fruit**, a bright orange capsule, about 2cm long, looks like a small apricot. It splits open in two to reveal dark orange-red seeds immersed in a sticky pulp.

Being drought- and frost-resistant, Weeping Pittosporum is an excellent hardy tree for planting in dry areas. Full sun and good drainage are its main requirements. The foliage is eaten by stock and is considered reasonably nutritious.

Hollywood *Pittosporum rhombifolium* to 15m

Usually found on the fringes of rainforests in near coastal situations north from Clarence River, NSW, to around Proserpine, Queensland, this tree has a dense canopy and a smooth, pale grey **bark**. The shiny, deep green **leaves** are alternate or clustered in groups of 3–5 at the ends of branches. They are ovate or mostly diamond-shaped, 6–11cm long, with toothed margins. It is sometimes known as the diamond-leaf laurel. Numerous small cream **flowers** are borne in dense terminal clusters in spring, but the beauty of this tree is the masses of small orange **fruit** held on the tree for many weeks in autumn and winter. This is an extremely popular ornamental street and park tree in warm districts.

Sweet Pittosporum *Pittosporum undulatum* 4–15m

In rainforests or moist gullies in wet sclerophyll forests, this tree develops a columnar trunk and dense rounded canopy. It readily naturalises new areas, where it is usually shorter with a more rounded spreading habit. Sweet Pittosporum is widespread, extending along the coast from southern Queensland to eastern Victoria and into north-western Tasmania. The thin, shiny, dark green ovate to lanceolate **leaves**, 6–15cm long and 50mm wide, often have wavy margins. Fragrant, creamy white, bell-shaped **flowers** are borne in terminal clusters in late winter and early spring. The **fruit**, an orange capsule about 12mm long, splits to reveal sticky brown seeds. This hardy ornamental plant is commonly cultivated.

Celery Wood *Polyscias elegans* to 10m

This small tree earns its common name from its fresh bark and young leaves, which are said to have the odour of celery. It often grows in rainforests or on the edge of rainforest clearings and has a wide distribution in eastern Australia from Jervis Bay, NSW, to northern Queensland and New Guinea.

It has a slender trunk with brown flaky **bark** that is often splotched with lichens. The bipinnate glossy green **leaves** are large, up to 10m or more in length, and have numerous, ovate to elliptic pointed leaflets, 5–12cm long and 3–6.5cm wide. They are aromatic when crushed. Small purple **flowers** are borne in a large terminal panicle beyond the foliage in autumn. The **fruit**, a rounded, slightly flattened berry, about 5mm long, is dark purple when it ripens in late autumn and winter. It is eaten by many native birds.

Young plants often have an umbrella-shaped crown of large leaves. This species is sometimes grown as a shady ornamental in warm temperate and tropical climates.

Umbrella Tree *Schefflera actinophylla* to 15m

This common street and garden tree in Queensland and NSW is known throughout the world as a hardy indoor potplant. In its rainforest habitat of northern Queensland and New Guinea, multiple erect trunks form a dense canopy of large compound **leaves**. Each 40cm-wide leaf consists of 7–16 light green glossy oblong leaflets that radiate on a long leaf stalk and resemble an umbrella.

Numerous clusters of small, bright red **flowers** are arranged on spectacular upright radiating spikes to 1m long (bottom photo). These appear near the top of the plant mostly in autumn and are followed by dark red fleshy **fruit**. Both the nectar-rich flowers and the fleshy fruit are eaten by native birds.

In warm to hot climates, the umbrella tree can be planted in a sunny position in the garden. Although easy to grow, a well-drained, humus-enriched soil with plenty of moisture during summer will give best results. In cool areas it needs protection from frost, but it can be grown in a container indoors in a reasonably well-lit position.

Alexandra Palm *Archontophoenix alexandrae* to 25m

This tall, elegant palm from northern Queensland rainforests tends to grow, often in dense colonies, along stream banks where there is a constant supply of moisture. It has a slender straight trunk that is enlarged at the base. The gracefully arching pinnate leaves, up to 3m long, have a large sheathing base encircling the trunk. They are dark green above and silvery grey on the undersurface, which readily distinguishes this species from the similar Bangalow Palm, *Archontophoenix cunninghamiana*. Old leaves are shed cleanly from the trunk, leaving ringed scars. Large panicles, to 40cm long, of small creamy white **flowers** emerge from the base of the leaf sheaths in summer and autumn. These are followed by rounded red **fruits**, about 1.5cm long, which are relished by birds and mammals, especially large fruit bats.

Alexandra Palm thrives, and is commonly cultivated, in tropical gardens and parks. It also adapts well to frost-free warm-temperate climates if given plenty of organic mulching and fertiliser and adequate water during dry periods. Indoor plants decline rapidly in dry conditions and grow best when given good light and ample moisture. Low temperatures should be avoided.

Bangalow Palm *Archontophoenix cunninghamiana* to 25m

This tall, erect palm is common from the south coast of NSW to near Bundaberg, Queensland, where it occurs mostly in subtropical rainforests or wet sclerophyll forests along stream banks or moist situations, often in small groves. The very slender trunk is slightly enlarged at the base and is marked by rings where the leaves have been shed. The arching pinnate **leaves**, about 3–4m long, have large sheathing leaf bases and numerous linear leaflets with long elongated tips. They are dark glossy green on both surfaces.

Branched pendulous panicles, up to 40cm long, of small, pale purple **flowers** appear beneath the bases of the leaf sheaths in summer and autumn. The spathes that house the unopened flowers were used by Aborigines as water containers. The bright red or orange globular **fruits**, about 1.5cm long, ripen in autumn and are eaten by a variety of birds and mammals.

In warm areas, the Bangalow Palm is extensively planted as an ornamental and grows rapidly, provided ample water is assured. It will tolerate cooler conditions than the Alexandra palm, but is frost-tender while young. This palm makes an excellent container plant for sheltered courtyards, patios or verandas.

Cabbage Tree Palm *Livistona australis* 2–30m

Widely distributed along the eastern coast of Australia to as far south as East Gippsland, Victoria, this tall, slender palm inhabits subtropical and temperate rainforests and sheltered eucalypt forests. It often grows in colonies in moist or seasonally flooded sites.

The grey trunk is rough and finely fibrous and shows the scars left by the shed leaves. The large stiff shiny **leaves**, to 1.5m across, are fan-shaped and divided into numerous drooping segments. They are carried on thick stalks, up to 2m long, which often have sharp curved prickles along their margins. Small yellow **flowers** are borne on a branched drooping panicle, about 1m long, in spring. The globular black **fruits** are about 2cm across.

In early days of European settlement, the cabbage tree palm was cut down for fencing and the young leaf midribs were once plaited into broad-brimmed hats. The terminal bud, or heart of palm, was eaten raw or roasted in hot ashes by Aborigines and early settlers but unfortunately, the removal of this bud kills the palm.

This is a stately ornamental palm for the garden or park, planted either singly or in groves. It prefers a moist, humus-enriched soil and some protection from strong sun and wind. It may also be grown as a tub specimen for many years and used as an indoor plant while it is young.

Weeping Cabbage Palm *Livistona decipiens* 10–15m

This species has a more northerly distribution than *Livistona australis*, occurring from Cooloola to round Townsville, Queensland, growing mainly at rainforest margins and in open eucalypt forests. It has a grey trunk and large, fan-shaped **leaves** about 1.5m long, with long trailing segments that create a beautiful soft, weeping crown in mature specimens. Small yellow **flowers** are borne in panicles, 1m long, in late winter and spring. These are followed by rounded black shiny **fruits**, to 2cm across. This is an outstanding ornamental palm for tropical, subtropical and warm temperate gardens. A sunny position, plenty of moisture and good drainage are its main requirements.

Sand Palm *Livistona humilis* 1–5m

This small, slender-stemmed palm grows mostly in open forests and woodlands, often forming dense stands beneath eucalypts. It is endemic to the NT where it is common in the Top End. The stiff, fan-shaped **leaves**, about 80cm across, are divided into numerous pointed segments. The leaf stalks have numerous short, sharp prickles along the edges. Dead leaves form a persistent skirt-like arrangement beneath the crown. Many small yellow **flowers** are borne on erect spikes, to 2m long, that arch well above the crown. The ovoid **fruits**, to 1.5cm long, are green ripening to purple-black. Aborigines eat the young shoots of this palm and the growing tip is used for medicinal purposes, including the treatment of respiratory infections.

Spiral Screw Pine *Pandanus spiralis* 5–10m

This small tree is widespread and common across tropical northern Australia and is a very important plant of multiple use to Aborigines of the NT. The large fruits are roasted and the kernels extracted and eaten; the crushed fresh fruits are used to make a beverage. The fleshy white part of young leaves is also eaten, as well as being used as a remedy for mouth and throat infections. A poultice is made from the growing tips and applied to relieve aching backs and swollen joints. The leaves are dried and stripped and woven into baskets, mats, dilly bags and rope.

The Spiral Screw Pine is a distinctive tree with a spiral pattern on the trunk formed by leaf scars (bottom photo). Prop-roots are absent. The stiff, spirally arranged strap-like **leaves**, 1–2m long, are crowded toward the ends of the branches. They have numerous small prickles along the margins and the back of the prominent mid-rib. Small creamy white **flowers** are borne in dense terminal spikes, 3–7cm long. These are followed by large rounded **fruits**, 15–20cm across, composed of numerous wedge-shaped woody segments. They are green at first and change to deep orange red when ripe.

This species grows in low-lying areas where the drainage is poor, often fringing lagoons, billabongs and freshwater streams, where it may form large, dense stands. It also sometimes grows along the beaches.

Coastal Screw Pine *Pandanus tectorius* 5m

Pandanus is a large, widely distributed genus of some 200 species found in tropical and subtropical regions of the world, often in coastal areas. *Pandanus* is from pandang, the Malay name for these plants. In parts of tropical Asia and the Pacific, some species are grown for their edible fruits and the scented leaves are used in cooking.

This species is fairly common along the coastal fringe from about Port Macquarie, NSW, to northern Queensland. It is a small tree with a rounded canopy supported by a series of thickened prop-roots at the lower part of the stem. The strap-like, prickly-edged green **leaves**, arranged spirally at the end of branches, are long and narrow. The short-lived, small creamy **flowers** are produced in dense clusters and followed by a large pineapple-like **fruit**, up to 25cm across, made up of many individual red or yellow woody segments, each of which contains seeds. The fleshy part of the ripe fruit is edible. The seeds, which are embedded in the woody segments, are also edible. They are small but are very high in protein and have a good nutty flavour.

The Coastal Screw Pine is resistant to salt spray and is a decorative small tree for warm seaside gardens. It requires full sun or partial shade and a well-drained soil. When young, it makes an attractive foliage plant for a container.

Glossary

Alternate. (Of leaves) arranged at different levels along a stem; not opposite
Anther. The pollen-producing top part of the stamen
Arboreal. Living or situated among trees
Aril. The fleshy outer covering of some seeds, often brightly coloured
Axil. The upper angle between a stem and leaf
Berry. A fleshy, many-seeded fruit with a soft outer portion
Bipinnate. (Of leaves) having leaflets growing in pairs on paired stems
Bloom. A thin layer of white, waxy powder on some stems, leaves and fruit
Bract. A small leaf-like structure that surrounds or encloses a flower or group of small individual flowers
Buttress. (Of roots) a flattened expansion of the lower part of the trunk and root in rainforest trees
Calyx. The outer series of floral leaves, each one a sepal
Canopy. The topmost layer of branches and foliage of a community of trees
Capsule. A dry fruit which, when mature, dries and splits open to release the seeds
Cone. A woody fruit of a conifer, made up of overlapping scales. Also applied to other woody fruits such as those of banksias and casuarinas
Corolla. All the petals of a flower
Crown. All the branches of a tree
Deciduous. Shedding leaves once a year
Drupe. A succulent fruit with a stone enclosing one or more seeds
Elliptic. (Of leaves) oval and flat, broadest across the middle and tapered equally at both ends
Emergent. Tall trees that tower well above the general forest canopy
Entire. (Of leaves) having smooth margins without teeth or division
Exserted. Protruding from surrounding parts, as when the valves project above the rim of a eucalypt capsule
Falcate (Of leaves) curved and tapered to a point like a sickle, as many eucalypt leaves
Follicle. A dry fruit that splits open along one side only and contains more than one seed
Frond. The (often divided) leaf of a fern, cycad or palm
Glabrous. Smooth and hairless
Gland. (Of plants) a liquid-secreting organ, usually on leaves, stems and flowers
Glaucous. Covered with bloom, often giving a greyish or powdery appearance
Habitat. The place or environment in which particular plants and animals normally live
Incurved. (Of leaves) curved or bent inwards or upwards
Inflorescence. The structure that carries the flowers, which may be arranged in a number of ways, such as a spike or umbel
Involucre. Leaf-like structure enclosing a flower
Kino. A dark, reddish resin-like substance developed in the veins

of bark or wood, especially common in eucalypts
Lanceolate. (Of leaves) shaped like the blade of a lance, usually broadest at the lowest half
Linear. (Of leaves) long and narrow with more or less parallel sides
Littoral. Growing near the shoreline
Mallee. Shrubby eucalypt with multiple stems arising from an underground rootstock known as a lignotuber
Margin. The edge or boundary line of an organ
Midrib. The main central vein that runs the full length of a leaf
Node. The point on a stem where one or more leaves arise
Oblanceolate. (Of leaves) lance-shaped, with the broadest part toward the tip
Obovate. (Of leaves) egg-shaped with the broadest part toward the tip
Opposite. (Of leaves) arising in pairs at the same level, but on either side of the stem
Orbicular. (Of leaves) circular, or almost so
Ovate. egg-shaped
Palmate. (Of leaves) divided into lobes or leaflets that radiate from the leaf stalk like the fingers on a hand
Panicle. An inflorescence with many branches, each of which bears two or more flowers
Peduncle. The common stem that supports a group of flowers or fruits
Peltate. (Of leaves) circular, with the stalk attached to the middle of the lower surface instead at the edge of the leaf
Pendent. Hanging down
Perianth. The calyx and corolla of a flower
Petiole. The stalk of a leaf
Phyllode. A flattened leaf stalk that functions as a leaf, as in many wattles
Pinna (Pl pinnae) – one of the primary divisions of a pinnate leaf (leaflet)
Pinnate. (Of leaves) a compound leaf, divided once with leaflets arranged on both sides of a common stalk
Pubescent. Covered with short soft downy hairs
Raceme. An inflorescence where a series of lateral flowers, each with a stalk, is arranged along a single stem
Receptacle. The enlarged uppermost part of the flower stalk on which the floral parts are borne
Sclerophyll. A group of plants with hard stiff leaves
Sclerophyll forest. A forested area dominated by trees with sclerophyll leaves, namely the eucalypts
Sepal. One of the separate parts of the calyx, usually green and leaf-like
Spathe. A large bract enclosing the inflorescence
Stamen. The male portion of a flower, comprised of a pollen-bearing anther and the supporting stalk (filament)
Stigma. The part of the carpel that receives the pollen, usually at the tip of the style
Sucker. A new shoot growing from the roots of the parent tree
Terete. (Of leaves) slender and cylindrical
Terminal. Situated at the tip
Trifoliate. Having three leaves or leaflets

Umbel. A cluster of individual flowers where several flower stalks arise from the same point
Valve. The segment into which some dry fruits separate when splitting open
Venation. (Of leaves) the arrangement or pattern of veins
Whorl. Ring of leaves or floral parts encircling a stem at the same level

Further Reading

Beadle, N.C.W., Evans, O.D. & Carolin, R.C., 1972, *Flora of the Sydney Region*, A.H. and A.W. Reed, Sydney
Boland, D.J. *et al.*, 1984, *Forest Trees of Australia*, (4th Edn), Thomas Nelson (Australia) Ltd., Melbourne and CSIRO Melbourne
Brooker, M.I.H. & Kleinig, D.A., 1983, *Field Guide to Eucalypts: Vol. 1. Southeastern Australia*, Inkata Press, Melbourne
Costermans, L.F., 1981, *Native Tress and Shrubs of South Eastern Australia,* Rigby, Adelaide
Elliot, W.R. & Jones, D.L., 1980–97, *Encyclopaedia of Australian Plants Suitable for Cultivation*, Vol. 1–7, Lothian Publishing Co., Melbourne
Harden, G.J. (ed), 1990–92, *Flora of New South Wales* Vol. 1–3. New South Wales University Press, Sydney
Harmer, J., 1975, *North Australian Plants* Part 1, Society for Growing Australian Plants, Sydney
Hill, K.D. & Johnson, L.A.S., "Systemic studies in the eucalypts: A revision of the bloodwoods, genus *Corymbia* (Myrtaceae)", in *Telopea* 6:(2–3), 185–469
Jones, D.L., 1984, *Palms in Australia*, Reed Books, Sydney
Wheeler, J.R. (ed), 1992, *Flora of the Kimberley Region*, Department of Conservation and Land Management, Western Australia
Williams, K.A.W., 1979 & 1984, *Native Plants of Queensland*, Vols 1 & 2, Williams, Brisbane
Willis, J.H., 1972, *A Handbook to Plants in Victoria*, Vol 2, Melbourne University Press, Melbourne

Index